The Shark and the Jellyfish

The Shark and

Vanderbilt University Press <small>NASHVILLE</small>

the Jellyfish

MORE STORIES

IN NATURAL HISTORY

By Stephen Daubert

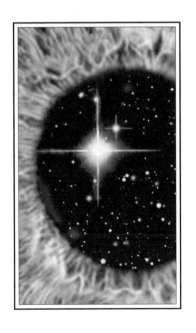 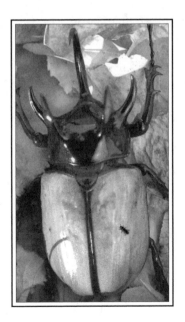 

13 12 11 10 09   1 2 3 4 5

This book is printed on acid-free paper made
from 30 percent post-consumer recycled content.
Manufactured in the United States of America

Jacket design: Cheryl Carrington
Text design: Dariel Mayer

Library of Congress Cataloging-in-Publication Data

Daubert, Stephen.
The shark and the jellyfish : more stories
in natural history / Stephen Daubert.
p. cm.
Includes bibliographical references and index.
ISBN 978-0-8265-1629-9 (cloth : alk. paper)
1. Natural history. I. Title.
QH45.2.D377 2009
508—dc22
2008029024

Contents

Preface **vii**

Part 1. Field and Stream

Accidental Airmail **3**

The Essence of Survival **9**

March of the Oaks **15**

Wolf Spring **22**

Where Nothing Grows **30**

Part 2. Air

Eye of the Needle **39**

Spider on the Fly **43**

Sky Walkers **49**

Nutcracker **56**

Flying Lessons **69**

Part 3. Sea and Shore

Sanderling **79**

Unseen Masters of the Sea **91**

Water World **101**

Sturgeon **110**

Life in the Sky **125**

Part 4. Forest

The Bitter Taste of Success   133

Fair-Weather Desert   138

Tree-Squirrel Fungus   144

Focal Point   154

Puppeteers   158

Part 5. Earth and Stars

The Light Fantastic   165

Gold   171

Incandescent Falls   177

Window on the Sky   183

Liaisons to a Rare Planetary Alignment   189

A Dangerous Place   198

Follow the Threads on the Web   207

Index   209

Credits   213

Preface

So many readers of *Threads from the Web of Life*—students, parents, science teachers, naturalists, and wildlife conservationists—have reacted to its stories with stories from their own experiences, that I decided to put together a sequel, a new set of adventures in the ecology of the natural world. I title the new book *The Shark and the Jellyfish*, after one of the twenty-six stories in it: like that one, many of the stories reveal surprising interconnections and boundary crossings.

Students of the history of the earth and the life upon it are natural storytellers. One of them may pick up a pebble from the trailside and describe its origin starting from the fires inside a dying star—where oxygen and silicon are produced by the fusion of helium atoms, then thrown into space, eventually coalescing into the rocks that form new planets. Another natural historian might look to the opposite side of the trail and begin a description of the DNA in a sapling there. That DNA encodes a record of the history of life on earth, read in the genes it shares with all other organisms. It also encodes the blueprints for the formation of cells, which form organs, which form organisms. This description of DNA will have been prelude to the story of one cell—a cell that divides into millions of daughters, which form into a sheet of tissue, which forms the autumn leaf now twirling round its stem between the storyteller's fingers. In the same way, a lone mushroom at the foot of an oak might prompt another naturalist to claim that the living landscape all around is one single being—the roots of every tree connect with all the other trees through a network of symbiotic fungi that links the entire forest together into a single, grand organism.

These storytellers would highlight spots in their scripts with points of fact we can all see, facts that anchor their stories to reality. At the same time they would call upon our imaginations to breathe life into features of the natural world that lie beyond our sight. We will never witness the conversion of helium to oxygen in the core of a dying star. We cannot inspect the nucleotide bases of DNA stacked one-by-one upon each other in their helices—their dimensions are smaller than the wavelengths of light with which we see what we believe. We will never witness the forest-wide breadth of the microscopic fungal network interconnecting all the trees beneath the trail—it lies hidden underground and crumbles to nothingness in our hands as we unearth even a small part of it.

Nevertheless, these concepts serve their storytellers well. They conjure a framework of understanding upon which we organize the things we *can* see. We see the rocks, the plants, the animals, but through them we *imagine* the motions of tectonic plates, the capture of photosynthetic sunlight, the evolution of species. That framework of understanding allows us to predict what we will find in times and places not yet seen.

Stories in this volume employ that device. They flow from what has been observed, to illustrate what we would predict. We have not sailed at thirty miles an hour thirty feet above the Tasman Sea at midnight along with the Neon Flying Squid. (See the second story in *Threads from the Web of Life*.) Nevertheless, we have enough information to envision that flight. Inference of such events draws upon our creativity—the descriptions are conjectural, predictions of that which has not yet been confirmed directly. Likewise, the illustrations in this volume are also extrapolations—works of creative nonfiction.

Other narratives we will never witness directly are told in the impulses passing through the minds of the animals with which we share the planet. We cannot know their thoughts; nonetheless, we can project what we know of them into tales told as if seen through their eyes, so to see their reactions to new situations. Stories of that sort are also contained in the pages that follow. Each account describes one thread from the broadest of our imaginary tapestries—the web of life.

These threads are the subject of the age-old discipline of natural history. It is one of the longest-established of the sciences and has been subdivided and renamed many times. Nevertheless, natural history is still a very active field. Our knowledge of its facets is expanding at the same exponential pace as is that of the more recent scientific disciplines. In the Science Notes sections that follow each story, the reader will see that about a third of the citations are no more than ten years old. We are still driven—more now than ever before—to deepen our appreciation of the world around us and to weave a framework of understanding around what we have found so far.

Part 1. Field and Stream

The hunted hunter. An osprey attack
saves one mayfly from the jaws of a trout.

Accidental Airmail

Dark branches spread against the predawn sky. Among them larger silhouettes—ospreys—perch in silence. Their heads swivel now and then—when something flies above the distant cliffs, or a ripple disturbs the lake surface below. These fish hawks have felt a subtle seasonal cue, and now their eyes pounce on every detail. There is a particular softness in the air that portends the appearance of mayflies.

News of the impending hatch has spread past the lakeshore—out beyond the aspen and pine at the margin of the valley-bottom meadow. Eagles perched on granite crags are watching from the distance. Their sight is keen enough to count the black-and-white fish hawks as the sun lights the treetops. The eagles will soon move into position to crash this party.

Below the water's surface the first members of the hatch are already emerging from concealment. Mayfly nymphs—small, dark swimmers with long legs and lacy gills edging their bodies—climb the strands of algae and swarm over the rocks and snags on the bottom. Their internal clocks have been reset by changes in the water temperature that predict a day above the waterline well suited to the nuptial flights of adult mayflies.

Most of these nymphs have long since been ready to emerge but have delayed. They were waiting for their final environmental cues while the slower growers among them caught up with the rest. Now the skin has begun to split along their backs, and they are moving up through the water column.

Their synchronous emergence will maximize their chances of finding mates. They will appear all at once, in numbers that will confront their predators with an overabundance of prey—leaving most of the insects to survive. They will molt quickly on the surface, sprouting wings to carry them away from the feeding frenzy their appearance will evoke. On the shelter of the shore there will be more time for a second molt, which will produce the final, clear-winged adults.

Trailing paired filaments longer than their bodies, the adults will be ready to take to the air the following morning—appearing as clouds of undulating motes that fill the sun gaps in the shoreline understory. They will drift out above the water, spinning higher, then dropping, finally mating.

Their brief adult lives will conclude as they spill millions of eggs back into the lake, assuring the continuation of their annual emergence spectacle.

The trout are the first to notice the hatch. The rising nymphs are invisible against the bottom, but when they move up through the water, they draw the attention of the fish. When their targets emerge into view backlit from above, the trout rise in pursuit. Where the surface begins to ripple and boil with feeding fish, the ospreys stir into action.

One osprey falls from her perch, converting height into speed as she dives for the water. She has singled out her quarry—one larger trout among the many that are rising all across the lake. But just as she begins her acceleration, the fish changes course with a flick of its tail and fades from view.

So the fish hawk flares her tail and levels out, gliding on her momentum. She is low enough now to see fine details—the dimples on the water that mark mayflies riding on the husks of their nymphal skins as they dry their new wings, and ripples spreading from the places where those mayflies disappeared as the trout struck.

The osprey wheels through a wide arc, and her turn brings another rising trout directly into her sights, even larger than her first choice. So she tucks back into a dive and builds upon the speed she carried from her initial descent. She is tightly focused as she plummets, accelerating as the fish enlarges in her eyes. When she hits the water, her foot will punch through the surface faster than the quickest trout could react.

The osprey hits her target at forty miles an hour, her speed carrying her completely under but for the very tips of her wings. Her curved talons are longer than those of any other raptor, and scales on the soles of her feet taper to thorns that work like the fine back teeth behind the canines of her claws—her grasp is as fast as the bite of a barracuda.

She emerges again before the splash has settled, holding her wings high. She beats against the surface and pulls her catch from the water, crashing through the swarms of mayflies as she works to gain height.

Her course across the lake takes her away from the open air. As she climbs toward a gap in the trees where the river crosses the shoreline, she flips the fish into the sky and catches it again. This move aligns the body of the trout with her flight—head forward to minimize the air resistance, just as the fish would align itself in the water to swim streamlined against the current.

A wall of trees eclipses the view back across the open water as the osprey follows a bend in the river. Nonetheless, she imagines an eagle has seen her.

Those pirates have twice the wing surface area she does and could close the gap quickly while she is slowed by the big trout she carries. So she takes a sharp turn up a tributary stream and flies through a narrower canyon of trees that shelters a rising, rocky creek.

Above the hiss of an approaching cataract, the osprey finally hears broad wingbeats closing the distance behind her. Ahead, the creek is disappearing beneath a ten-foot cascade of white—a waterfall descending over a mossy rampart. She will lose speed as she rises to clear it, and the eagle coming on from behind will catch her.

This osprey weighs only four pounds, and the fish she carries hinders her agility. She cuts close to the trees—the catkins rippling in her wake as she skims the crest of the falls. That maneuver leaves her low over the lake that opens out beyond the crest—with her pursuer now upon her.

The eagle screams and jostles her, his great wings generating blasts of air pressure pushed ahead by his motion. She turns sharply away, nearly losing control of her airborne stability—she has to pass the fish between talons to maintain her center of gravity as she banks. But as she slides sideways through the air, switching the fish from side to side, she loses her grip and her catch slips from her grasp.

The eagle pushes close past her, and she inadvertently slaps him as she fights to regain her altitude. The impact sets the osprey spinning and causes the bigger bird to lose sight of his prize for a moment. The eagle recovers to see that it is too late to seize this fish from midair. He times his dive to pin it against the surface, but the fish jackknifes to the side as it slaps down and is gone in a flash.

The trout has landed in a lake that it could otherwise never have reached—isolated from the rest of the drainage in the valley by a waterfall impossible for a trout to scale. Its gills reinflate at once, but they are badly wind-burned, and so the water breathing is labored. The fish fades to a dull shade of gray and sinks into the depths. It will sulk in the shadows for a week, motionless, not feeding, slowly healing the wounds inflicted by the osprey's talons.

Eventually its stiffness eases, its appetite reawakens, and the fish moves out into its new domain. It discovers the rocky channel swept clean by an inflow current that tastes of snowmelt and fallen pine needles—and carries an occasional ant. Worms and beetle larvae cling in the crevasses between the rocks.

The trout avoids the exposure of the warmer shallows through which he can see the trees and sky. But where the undergrowth or the boulders and tree roots hang out over calm water, he finds water boatmen and mosquito wrigglers. Diving beetles rise around him, each carrying an air bubble on its abdomen.

The lake seems empty of fish, as evidenced by its bounty of trout forage. The deep center is a green world alive with subtle motion. The trout holds his position against the sunken arms of a great fallen cedar and watches as a menagerie of nymphal-stage insects works the substrate; they include larval dragonflies and damselflies, and caddis fly nymphs ensconced within the masonry of cylindrical cases they have fitted together grain by grain.

The fish's final discovery is of one more trout—a full-grown, healthy specimen—olive blue on top fading to iridescent pink at the midline, grading to silver below, all overlain with a pattern of circles and dots spreading out across its fins. The geometric texture of its scales is perfect but for a mar on one flank—a fault line in the pattern where an old talon wound has healed. The two fish drift together, each watching the body language of the other.

After the catkins have fallen and the season matured, the osprey sits again on a branch overlooking the lake behind the falls. Her bright yellow eyes catch everything—from a small hatch of stoneflies under willows hanging out over the surface, to the crayfish prospecting on the open silt twenty feet from the bank. The still water far from shore momentarily boils where a pair of trout rises to the surface. One of them leaps to strike at a mayfly drifting above; then the two fish fade from view, and the calm returns. In future years those fish will have built a population of trout in this lake that the osprey and her descendants can profitably forage. She glances skyward against the glare of the midmorning sun.

A mile above, a speck invisible from the ground turns through a wide, lazy circle. A map of the landscape spreads out beneath it—from snowy crests on the far horizon back toward the throat of a steep canyon that widens to become the valley directly below. The forest rises to draw a tree line halfway up the slopes; clouds tracing the northwestern skyline promise a storm for late afternoon.

Straight down, the trails and streams, ducks on the water, elk on the meadows—all are censused in the eagle's aerial survey. To their east, hidden ponds and creeks stand out through the dense forest, their waters shining golden through the branches in reflected sunlight. The tableau wheels beneath the big bird, his head swiveling to counter his own motion when something of interest comes into view. Even at this distance, he commands the smallest details.

The largest details of the flow of the landscape, however, are beyond the eagle's purview. The valley simultaneously rises up and erodes away at a pace that exceeds the time scales perceived by individual birds or fish. Brooks meander across meadows over the decades; lakes spread behind rock-fall dams, fill in with the sediments brought down from the steeper slopes, grow reeds, then willows, then aspen, and finally disappear under stands of lodgepole pine. Creeks back up behind obstacles until they overflow upstream to find other courses, leaving the original beds dry. Smaller streams capture bigger streams, diverting them across the slopes to pond in mountain meadows that eventually overflow from cliffsides—redirecting the drainage from canyons over waterfalls. Secondary seeps sheet across the sheer walls, growing stronger until they refill the original channels.

And while new streams and lakes may come into existence bereft of swimming things, they all soon grow trout. Even a watercourse interrupted by impassable cataracts or isolated from other drainages will still have fish moving in its depths. Natural populations of trout find their way into crater lakes, into spring-fed brooks that empty over steep shallow spillways, into pothole lakes fed by snowmelt. How do they do that?

The trout, the eagle, and the osprey spread their generations across this domain in the course of their coexistence. Ospreys carry their live prey for miles, over ridgelines that divide drainages, to perches by the shores of isolated lakes and streams. The eagles seek out and pursue them, winning a significant proportion of their calories through robberies. The constant contest between the two raptors occasionally results in the arrival by airmail of a trout that was about to lay eggs when it was snatched and then was dropped, sometimes into an empty stream. As a result, its eggs will hatch in a waterway that was free of fish but is well suited to support them—their insect prey having flown in and established themselves beforehand.

Such stream-stocking events may be rare on the time scales of the lives of fish or eagles or ospreys, but they replay again and again over the geologic time scale on which the landscape evolves. The chance trout finds itself planted in a new lake or stream often enough that all the waters are inhabited almost all the time—to the benefit of the populations of all three species.

Science Notes

This tale states a hypothesis for the dissemination of fish among waterways: that they are carried and spread by chance by birds of prey. An osprey may drop the fish it has carried for miles when harassed by an eagle. It may also drop a fish without provocation; for example, an osprey carrying heavy prey over long distances may be dragged down until it drops its fish due to fatigue (MacDonald and Seymour, 1994). Eagles attacked by other eagles drop fish. The extent to which fish dropping by osprey and eagles currently contributes to the stocking of isolated streams and lakes remains to be established. One complication to such a study at this time would be the larger contribution to that effort by human aerial stocking programs.

Ospreys have pointed scales on the soles of their feet (Poole et al., 2002). Their talons are longer in relation to the size of their feet than are those of other raptors. Eagles may gain 10–20 percent of their diets by kleptoparasitic attack on other birds (Brockman and Barnard, 1979).

The first molt of the adult mayfly produces a subimago that flies from the water. A second molt ensues to produce the adult imago. Mayflies are the only insects that molt again after achieving winged form. There are two thousand species of mayfly, each with its own strategy for synchrony, hatch date and time of day, and adult life span. Mayflies are one of the most successful insect lineages. They have witnessed more geologic change than any other creature on the land, having become established during the Carboniferous epoch five hundred million years ago. The mountain ranges from which their streams once issued have risen and crumbled to sand time and time again.

References

Brockman, H. J., and C. J. Barnard. 1979. Kleptoparasitism in birds. *Animal Behavior* 27:487–514.

Knopp, M., and R. Cormier. 1997. *Mayflies: An angler's study of trout water Ephemeroptera*. Helena, Mont.: Greycliff.

MacDonald, J., and N. R. Seymour. 1994. Bald eagle attacks osprey. *Journal of Raptor Research* 28:122.

Poole, A. F., et al. 2002. Osprey: *Pandion haliaetus*. In A. Poole and F. Gill, eds., *The birds of North America*, no. 683. Philadelphia: Academy of Natural Sciences; Washington, D.C.: American Ornithologists' Union.

The Essence of Survival

A myriad of scents are borne to the rabbit—the spices of the foliage and perfumes of the blooms, the musks of the animals, the ferments of the soil. They describe the landscape in detail more intimate than is available through any of his other senses. Those aromas fade and evolve while they travel on the air. Their characteristic flavors depend on their concentrations, so as they diffuse and thin out, each successive dilution smells different—they change as the distance from the sources grows.

These essences fade the same way at their sources—with time. The redolence of the flowers, the scat, or the scent marks of passing animals change as the reservoirs of their aromas gradually evaporate away.

The rabbit had paid close attention to all these cues through the years. He confirmed the details again every day while he made his rounds under the cover of the low scrub. By now he could recognize the distance to—and the age of—every airborne evocation of plant or animal life that floated his way, miles in any direction. Accumulated, unchanging scents told him he was becalmed—under a windbreak, or under a windless sky. His sense of smell told him about the weather. His nose was as critical to his quest for food and shelter as his tall ears were to his safety.

As he moved through his territory, his meticulous attention to the world passing beneath his feet showed no hint of his quickness and potential burst of speed. He was calm and measured as he checked the objects protruding into his path. Most of them carried his most reassuring fragrance—they smelled like him. He scent marked them all as he passed. His own scent was a distinct mixture of substances with different volatilities. It aged predictably—when it was fresh, its essence was dominated by the molecules that evaporated most rapidly; when it was old, the most volatile molecules were gone, leaving faint remainders dominated by different, less volatile components. He knew when he was passing beyond the usual borders of his territory because his scent marks smelled older.

He stopped and bit into a leaf, ignoring the bitter taste while he tested the scents that floated from the crushed sample into his nose. He gnawed through a dead twig—a compulsive habit that wore down his teeth and prevented them from growing out too far. Then he marked the freshly exposed wood by rubbing it with the scent glands under his chin.

No thicket was so impenetrable that the rabbit could not tell what happened within and beyond it. Airborne clues filtered to him through the maze of branches. The incense of indole hung in a tangle of chamise stems—an indication that the primrose was coming into bloom nearby. The aroma of freshly turned earth beside the path was mixed with a trace of musk, a sign that some insectivorous animal—a weasel, maybe a skunk— was visiting his territory.

Farther on he recognized a jot of quinone at the base of a dead shrub— a black beetle had assumed its defensive handstand here just today. Head down, tail held high, it had sprayed its repellent mace at this spot. The effect was still sharp and bitter—too strong to get a nose close to. But already the rabbit could sense it changing, altering its flavor with the passage of the hours.

The aromatic galaxy surrounded him with a sense of normality and a measure of the seasonality that circumscribed his life. Nonetheless, most of those cues were of only secondary interest. He stopped abruptly and backtracked one step, brought his face down to a whisker's length above the ground, and sniffed a little round bump embedded in the earth. It turned out to be just an old rabbit pellet, and not from a different rabbit, but his own. His interest in the possibility that another of his kind might have visited the area passed, and he resumed his rounds through the brush. What he sought most avidly was the scent of a female rabbit. That was the rarest of discoveries—he had not found one since the coyotes had come to the area long months ago.

The pair of coyotes could tell from the scent of the earth that this was still active rabbit territory—even though they had long since hunted out most of the rabbit here. They were as perceptive of the signs borne on the air as were their quarry.

They detected a faint hint of skunk on the breeze. Sometimes they hunted skunk, or badger, or porcupine, but not often; these heavily defended prey required so much effort to stalk and ambush successfully that the coyotes usually just avoided them. The coyotes ate mice, insects, and carrion; they ate grass and berries. They would prefer rabbit if they could find it, but they hadn't found much lately.

Coyotes have very long memories. The lead female had noticed the weather changing—the breeze coming around to the north, the light failing prematurely, the color draining from the landscape. Now, as the afternoon lengthened, she was searching through the brush for a low rock outcrop she recalled. The hollow in its lee would offer shelter to an animal exposed to the freshening wind. The male, younger, followed along obediently. His experience was that she often led him to food.

The earth was an evolving tapestry of fragrances, a record going back years. The rabbit never went far before his nose pulled his head down to follow the newest threads. He navigated between old scent marks of other creatures—oils and musks and urines, complex mixes of lymph and microbial ferment and pheromones—the individuality of each dulled with exposure to the sun's rays and the dry air. He was able to separate the scent marks from one another and from the smell of shed leaves, buds, and husks crushing into the ground. Places where animals had died provided particularly stable landmarks. The land is covered with their overlapping echoes from across the cycles of seasons—pungent initially, ever fainter day by day.

The rabbit found an arresting mix of such signs on a spot beneath the brush, so he stopped and dug into the earth. Between bouts of digging he thrust his nose into the hole, and the stories bloomed out of the ground—the more volatile essences that had evaporated from the crust were still there farther down, waiting for him to experience again. Then his head popped up, and he froze for a moment—one ear forward, one ear back—motionless but for the twitching of his nose held to the wind. He had to check that his work at this distraction had not masked telltale scents or sounds floating in from the middle distance.

As he heightened his attention, the rabbit became aware of a change stealing over the landscape. The aromas of the darkening afternoon had intensified, now nearly overloading his senses. The temperature had fallen and the humidity had risen, conspiring to minimize evaporative loss from the leaves. The plants had responded by opening their pores. The soft breeze had become freighted with the camphor zest of sage that poured from the foliage nearby, mixing with the tang of menthol. He sat up straight and pulled his front legs back down across his chest, lining up all twenty toes in a row, finally moving nothing but his nose. This outpouring of spice indicated one thing to the rabbit—rain.

So he bent his course, moving off with his usual deliberation, stopping every so often, his padded furry footfalls silent. Soon the rustling of the higher branches in the breeze ceased, and the first soft drops began to fall in the silence, big drops ticking down here and there against the dry leaf litter.

He found himself drawn to a particular hollow beside a rock, but before he stepped down into its shelter he paused. The raindrops were masking all other sounds, and when he moved past the edge, the noise of half those drops disappeared—the half that were eclipsed behind the wall beside him, which was just taller than the tips of his ears.

The rabbit hesitated above the hollow. With only a few minutes of light remaining, darkness would soon preclude his search for other shelter. He stood, listened, and tasted the air.

Night was closing in. The rain plastered the coyotes' fur down across their faces. They were stopped, looking at a low rock protruding from the ground ahead, dimly visible before them in the gathering darkness. The second coyote inferred from the stance of the first that their quarry would be behind that edge, so he moved left when she went right—they separated and approached the objective from both sides. Their advance was careful, softened by the dampness absorbing into the soil and masked by the rainfall.

The lead coyote stopped just before she rounded the rocky edge and gathered herself to spring upon the small hollow. Though he saw no shadow or motion beyond that edge, the male took his cue from the female's lead, coiling himself to spring when she did.

The coyotes pounced together, their paws pinning down the earth in the lee of the rock. There was nothing there. The female stood back on her heels, quivering in anticipation, ears held high, eyes flicking side to side, ready to spring at any motion, but the area was deserted. The male stood tense, ready to follow her, but there was nothing to chase; he looked sidelong at the female, still holding at the ready.

Suddenly the female's head spun around. There was movement in the middle distance—a small creature retreating through the darkness from the area of their attack. She shot off in pursuit, the male close behind. They quickly closed the gap as their quarry disappeared below a low rise, and the female leaped from the high point to dive upon her victim.

The coyote seemed to feel herself stop in midair as she finally visualized her target and tried to reverse her flight. A small, dark animal was posed head down before her, handstanding on its forefeet, tail held high.

At first the coyote only vaguely perceived the drops of warm liquid mixing with the rain on her face. But in the next instant the stunning, sharp sting of the skunk's mace crashed against her senses.

She rebounded from the assault as though she had run full speed into a wall. She could barely breathe or open her eyes against the overpowering, disorienting reek that clung to her muzzle. Through the commotion of her own stumbling recoil she heard the pounding of her partner's feet as he raced away into the night.

The stench would take hours to dissipate to less painful levels in her nose, and days more to fade enough that it did not render her sense of smell useless. She would be blind to the olfactory cues around her, and at the same time she would be easy for prey to avoid, and for the wolves to track—she would have to head out into the barren scablands to hide. The coyote let out a howl of pain and despair, frightening the skunk, who let fly with a second volley of musk.

On the edge of a clearing a large bush shuddered in the rain gusting from the south. In its lee a smaller bush protruding out into the open stood unmoved by the storm. And in the shelter of that smaller overhang an unobtrusive lump in the darkness settled undetected toward sleep.

The rabbit had left the rocky outcrop and continued moving through the brush until it was too dark for him to see, or be seen; he stopped where the night overtook him. Now he was hidden from sight by darkness, and hidden in sound behind the endless patter of the falling rain. All around him the scents from his dominion were dissolving, washed down from the branches to flavor the rivulets that flowed away into the soil. The rain would wash the odors from rabbit fur more thoroughly than he could by grooming; soon he would be invisible to predators who hunt by their sense of smell.

The rain had the opposite effect on the other scents in the world around him—it brought things into sharp focus. It mobilized aromas from deep within the earth, so they could diffuse back to the surface where he could again detect them. It freshened his forage plants; their pores opened wide, and the particular bouquets of each filled the air becalmed by their branches.

The rabbit's nose worked now and then, distracting him from sleep, its revelations undiminished by the raining blackness without. A lone coyote howled in the distance, but the rabbit did not flinch. He never moved in response to a sudden sound, no matter the proximity—movement could draw notice. The call went unanswered—there was no chorus of coyotes in the area—and soon the curtains of rain closed again over all his senses.

The rabbit tucked his feet back under his breast and closed his eyes. He folded his ears down against his back, keeping them dry inside, deafening the impact of big drops all around. The water percolated down through his fur, forming a layer next to his skin warmed by his body heat and insulated by the fur above it; the ground beneath him warmed as well.

The steady rain would persist while the storm front passed. Thunder-showers would announce its departure by morning. Then he would find his scent garden transformed into a new tapestry to explore all over again.

Science Notes

Skunks, weasels, and badgers are in family Mustelidae—named after their must (musk) glands. Family Mustelidae, like family Canidae (which includes the coyotes), is in order Carnivora. Porcupines are rodents, an order closely related to order Lagomorpha (rabbits).

Menthol, quinone, and indole (see diagram, left to right) are examples of the thousands of small organic compounds that mediate the metabolisms of Earth's life forms. These molecules are built around hexagonal rings of carbon atoms. Line segments represent bonds between atoms in the carbon skeleton diagrams; by convention, carbon-bound hydrogen atoms are not shown. These compounds are not very water soluble and therefore evaporate readily. Indole offers an example of the metamorphosis of aromas with dilution—its concentrated form has a decidedly fecal smell, while at concentrations in air a thousandfold less, it is a common component in the scents of flowers (Dudareva and Pichersky, 2006). Menthol is a characteristic volatile in the foliage of plants in family Labiatae (mints); 1,4 benzo-quinone (hydroquinone is shown at center in the diagram) is a component of the mace sprayed by the bombardier beetle (Eisner et al., 2000). The odor of skunk is dominated by mercaptans—organo-sulfur compounds. Skunks and bombardier beetles assume a similar posture—posterior held high—when threatening to spray their defensive maces.

References

Dudareva, N., and E. Pichersky, eds. 2006. *Biology of floral scent*. Boca Raton: CRC Press.

Eisner, T., et al. 2000. Spray mechanism of the most primitive bombardier beetle (*Metrius contractus*). *Journal of Experimental Biology* 203:1265–75.

March of the Oaks

Tracks—rattlesnake, roadrunner—meander undisturbed beneath the branches, here on the chaparral ridge. The silence is punctuated only by occasional silverbush lupine seed heads exploding in the heat. Spring wildflowers promised by those seeds will wait buried in the dust for months to come.

But on the highest arch of the ridge, a little piece of parkland flourishes in the middle of hundreds of square miles of dormant buckthorn. A coppice of cool, leafy trees draws a stripe of shade across the sun-struck sea of shoulder-high scrub brush. The grove is thirty feet tall and fifty yards long, and rustles in a breeze felt nowhere else around. This oasis of green appears to have been transplanted to this spot from some far-off black oak forest.

Did a scrub jay carry a single acorn here long ago, flying across the dry country to plant a founder tree? No, this is more than oaks—it is a rich community of plants. Coffeeberry grows in the understory miles from its usual habitat, along with greenleaf manzanita, gooseberry, wild peony, and poison oak. Nashville warblers—birds never found in the low chaparral—sing from the canopy as confidently as they would in the distant forested foothills riding off to the north, the dark expanse of which is now obscured from view by the midday haze.

This woodlot is not lost in the landscape—it is lost in time. These trees once spread much farther. Tough scrub oak and chamise chaparral dominate the slopes now, but where they fall into the shade under the canopy of black oaks, the whiptail lizards of the barren rocks give way to skinks slithering between woodland ferns, mosses, and sapling trees, their leaves still bronzed with a blush of seedling color. The entire wet forest cohort is thriving here in this isolated little plot, waiting for the next ice age.

The cool oasis provides a welcome respite from the glare of July. The vista seen from its shade seems to reach forever—parallel ridges all covered in the same dark fire chaparral rise to eye level and recede away one after the other. Intervening hazes render each silhouette successively less distinct as the ridges retreat toward their final barrier—the shore of the North Pacific Ocean fifty miles to the west. Snowfields crown the mountains rising on the other side of the broad valley below to the east. The hard blue sky lightens as it descends to the south—fine dust rises into the atmosphere there from the distant southern desert. In the opposite direction the sky does not lighten but stays sharp—reflecting the icy peaks below the far northern horizon.

It is no coincidence that this island of woods is perched on the most picturesque vantage for miles around. It stands just here because this is the rain crest—the highest pitch on the brow of the ridge, the place that reaches closest to the overcasts of the rainy season. The topography here deflects the clouds upward to colder heights as they pass over, wringing rain from the moist air.

This particular peak gets twice as much rain as the slopes under the rain shadow to its lee. As much rain falls now on this spot as fell everywhere over the province ten thousand years ago—when the glaciers to the north were at their closest. At that time, the cool oak forest extended unbroken far to the south to lands that are now desert.

The sky island lives here as it did long ago, thriving as though the last ice age had never ended. It still rains here as it did then, before the rest of the forest on the slopes all around gave way to dryland flora. This relic stand of oaks provides a refugium for the assemblage of plants and animals that were everywhere when black oak covered these hills as thickly as the chaparral covers them now.

Pacific black oak is a transition species, living on the weather boundary between northern conifer forest and dry chaparral, a boundary always on the move across geologic time. The southernmost margin of black oak forest advances with the expansion of the ice age glaciers. That margin moves behind a blue wave of shooting stars and phacelias that splashes a new flora across a landscape shifting to cooler winters and wetter summers. And when the cycle swings and the glaciers retreat, a different, southerly wave moves back over the ground, flowing with warm shades of buckwheat and lupine, leading a trend toward milder winters and summer drought

Shoots spring quickly from black oak roots when everything aboveground is laid waste by the inevitable, fifty-year firestorm moving in from the chaparral to the south. The adult trees will not tolerate shade from Ponderosa pine or incense cedar encroaching from the north, but black oak seedlings are shade tolerant and will push up through the undergrowth. These traits, not shared by the conifers, give this tree the advantage at the

Oak spring. Pacific black oak seedlings
emerge into a ridge-top habitat.

forest's edges. But in time, the oaks are replaced by pines if the cool climate persists, and by the scrub if the influence of the desert returns.

Now, as the endless summer lingers over Black Oak Ridge, the advancing desert boundary is close at hand. The bark splits on the last of the old white alders in the canyon below. These trees are victims of a recent string of early summers that has left their draws without running water beyond April. The warming trend in the current interglacial period has also killed manzanita trees, sending their trunks rolling down rocky slopes on the ridge's southern extreme. The springs in the rocks these full-sized specimens depended upon have ceased to flow; now, their peeling iron-red skin is fading to a papery gray, waiting to be vaporized by the next wildfire.

This region where the trees fail to reestablish from one generation to the next is the moving edge—the site of the retreating forest margin. Far to the north the ice is receding, pulling the southern border of the cool forest with it along the ridge, each variety of plant disappearing at its own pace. Fallen boles of cottonwoods lie across dry washes down the hill, in canyons that were once wet enough to support their growth. These big trees are the last of their kind there—the seedlings of their next generations have withered and died to the last in recent decades, in the dusty Augusts at the end of their first summers.

As their favored conditions desert them, the narrowly adapted trees and flowers linger longest in those exposures that retain their requisite conditions. On the far southern ramparts of the ridge, flora that is common in the cooler climate of Canada can still be found holding on in sheltered, north-facing hollows that provide the mild, moist mornings now rare elsewhere on the ridge. On dry, south-facing slopes farther north, plants of the desert foothills appear occasionally on well-drained soils. And on the ridge crest, the oaks survive in their sky island.

Most of the tree species that have ever grown on the Earth are now gone. Their survivable range eventually contracted down to one last refuge, then continued to contract, squeezing the last holdouts into extinction. We have discovered a few of these final refugia and reversed that course for some. The last of the wild ginkgo trees died out in the mountains of China during the past several millennia. The few cultivated examples that survived owed their existence to Buddhist monks living in monasteries there. The final population of dawn redwood was found near those same retreats in the 1940s. Both species were rescued from oblivion by our incorporation of their last survivors into the human landscape.

The last stand of Wollemi pines was found in a deep, narrow canyon in the Blue Mountains of New South Wales, Australia, in the 1990s and is now being reestablished as well. Other extant species—redwoods, American chestnut, Monterey cypress, Torrey pine—find their ranges retreating, in keeping with the trend toward extinction. Ninety-nine percent of the species of plants—and animals—that have ever existed have disappeared.

The total number of species that exist at any one time, however, shows the opposite trend. More species are alive now than at any time in the history of the planet. Every time the moving boundary sweeps across the ridge, it carries more species than ever before. The increase in species with time is a consequence of the rhythm of the living tide that washes over the landscape on alternating waves of climatic warmth and chill. When the species of the migrating forest retreat, they leave pockets of their kind behind, leading to the isolation of some in refugia. The acclimation of stranded flora to the changing conditions of their confines results in the splitting of species, and the eventual generation of new ones.

Niches are divided, then subdivided, populating ever finer and more narrowly defined living spaces. There are now more than 150 species of phacelia, more than 180 species of buckwheat, and half again as many subspecies of each—varieties that are the seed stock for the eventual generation of even more new species. There are now more than 250 species of lupine—found from the melt-water footings of the glaciers to the desert sand dunes. Initially, there was only one.

Biological landscapes diversify as they adapt to changing conditions. The more diversity a landscape contains, the more resilient and stable it is. The tide of ice will continue its ebb and flow from the north, following the regular changes in the Earth's orbit—cycles lasting tens of thousands of years. And the plants will continue to respond to the challenges.

With the return of the glaciers, the southernmost refugia of black oak will be well positioned to expand. The sites where oak seedlings begin to survive in the sun to the south of their parents will mark the advancing forest boundary. That vanguard generation of seedlings will lead the march of the cool forest down from the ridges, through the cooler valleys, and then up through the chaparral on the hillsides. The shady margins will merge to surround and strand pockets of warmth-loving species in their own islands on south-facing slopes. The conifers will do the same to the oaks as they advance not far behind. The biological landscape shaped by these cycles is the one best suited to maintain itself—and the earth, air, and water it regenerates—against whatever comes: pestilence and parasites, floods and drought, fire and ice.

Science Notes

The low mountain and foothill provinces of the American Southwest and West are among the most floristically diverse regions in the world. This diversity is a legacy of the repeated march of climatic boundaries to the south, then back to the north, over recent geologic time (Chaney, 1938). Pacific black oak (*Quercus kelloggii*) is a prominent species in this province. (It is distinct from eastern black oak (*Quercus velutina*), which pursues a different but no less involved life strategy in eastern North America.) The dynamic range of Pacific black oak and its attendant flora is complex, driven by many forces (Chaney, 1938; Delcourt and Delcourt, 1987); its niche in dryland or conifer transitional successions is not completely understood. Without disturbance (e.g., by fire or human intervention), Pacific black oak is eventually crowded out of the best sites and remains only as a remnant in mixed conifer forest. Its decline may have been continuous since the retreat of the last ice age began sixteen thousand years ago. Increasing aridity is the probable cause for the shrinking range of the modern population (McDonald, 1990). Effects of that aridity are manifest in the ranges of all members of the ecotone, the boundary between forest and shrub zones (Breshears et al., 2005).

Relict groves of Pacific black oak are often of uniform height, a consequence of synchronous resprouting after the trees are destroyed by fire. The advance of oaks over geologic time is borne on the wings of acorn-carrying blue jays (Johnson and Webb, 1989). Associated black oak understory species include deer brush (*Ceanothus integerrimus*), green leaf (*Arctostaphylos patula*), and white leaf (*A. viscida*) manzanita. (The type member manzanita (*Arctostaphylos manzanita*) is more a full-sun, chaparral species.) Species counts listed here for the flowers are from Munz (1973) and may be underestimates. The glacial cycles, lasting tens of thousands of years, are synchronous with (and so possibly caused by) excursions of the ellipticity, axial tilt, and precession of the Earth's orbit, as shown by the calculations of Milankovitch (Berger et al., 1984). The fossil record shows examples of such swings in the Earth's climate that have seen warmth-adapted plants extend their ranges 1,500 kilometers to the north across the continental United States over a period of only ten to twenty thousand years (Wing et al., 2005).

References

Berger, A. L., et al., eds. 1984. *Milankovitch and climate: Understanding the response to astronomical forcing.* Dordrecht: Reidel.

Breshears, D. D., et al. 2005. Regional vegetation die-off in response to global-change-type drought. *Proceedings of the National Academy of Sciences U. S. A.* 102:15144–48.

Chaney, R. W. 1938. Paleoecological interpretation of Cenozoic plants in western North America. *Botanical Review* 4:371–98.

Delcourt, P. A., and H. R. Delcourt. 1987. *Long-term forest dynamics of the temperate zone.* New York: Springer-Verlag.

Johnson, W. C., and T. Webb. 1989. The role of blue jays in the post-glacial dispersal of fagaceous trees in Eastern North America. *Journal of Biogeography* 16:561–71.

McDonald, P. M. 1990. *Quercus kelloggii*: California Black Oak. In R. M. Burns and B. H. Honkala, eds., *Silvics of North America. Agricultural Handbook #654*, vol. 2. Washington, D.C.: U.S. Department of Agriculture.

Munz, P. A. 1973. *A California flora.* Berkeley and Los Angeles: University of California Press.

Wing, S. L., et al. 2005. Transient floral change and rapid global warming at the Paleocene-Eocene boundary. *Science* 310:993–96.

Wolf Spring

A brook cascades over the pile of branches in its path, following the steepest course in its race downstream. The sound of a lively spillway carries the promise of water aplenty for spring growth—and prosperity for the forest. For the beaver, however, it is a warning that his dam has suffered a breach. He is driven to stop what he was doing and staunch that sound wherever he finds it; it must be reduced to the thin trickling of many smaller leaks— and then to silence.

This break in the edifice of his modest waterworks was caused by the clumsiness of a moose. He had watched her feet punching through the surface, each step closer than the last as she ambled downstream through the center of his pond. She took her time, searching for a few thin reeds that other grazers might have missed. The dam was neither high nor wide, but instead of stepping over it, the beast had simply planted a foot right on top. Then she levered her frame across the barrier, and her heavy hoof crushed the structure beneath it.

The beaver worked on the low point as best he could, though he did not have the material to patch it. His stream ran through a forest that had grown sparse—the trees were scattered, with too much open space for him to feel secure venturing overland in search of lumber. So he cannibalized other parts of the structure, stealing sticks from here to fix a leak there, then turning around to discover that another leak had sprung at the site of the theft. He worked all day, making many small leaks out of one large one, struggling to oppose the irresistible tendency of water to find its own level, reducing the sounds as best he could to create in the end a dam that was a little lower.

He paused frequently to test the air, distracted by distant movements through the trees. Passing elk were common in the area. For as long as he had been here, the grazers had been increasing across the landscape. Their population had risen to its maximum limit. Browse was thin, hunger common, accompanied by social conflict among the hoofed hordes. The scene was set for a population crash. The only question was, What form would it take? Famine? Disease? Or the advent of a predator.

The grazers had left the woods barren. Small trees the beaver felled were often already wilted, stripped of their bark to a height equal to that of a deer standing on its hind legs. There was little understory brush to cover him while away from the water. What branches he had harvested were cached in the middle of his pond, a sunken store of bark food that was running thin, here at the end of winter.

The beaver worked on his modest waterworks with the help of just one other, a tenant beaver who did not share his lodge but lived in a hole in the bank—accessed through an underwater entrance. The first beaver tolerated the second in his pond in part because he recognized him by scent as his brother. The other beaver had migrated from a broken dam washed out by the flood surge from a cloudburst farther up the mountain the previous fall.

Only a few days later, as he worked submerged, arranging stones at the base of the dam, the beaver felt the shock passing through his chest of a tail-slap on the water. He spun away from his project, avoiding the surface altogether, and dove for the depths of the drowned central streambed. He flew twisting and banking through the turns of a path he knew well there, arriving in seconds at the submerged entrance to the lodge at the center of the pond.

As he emerged dripping into the first chamber of his fortress, he sensed another presence. Creeping up into the dry living room, he found his pond mate already inside, cowering against the far wall. He had long been patient with this unfortunate tenant in his domain but would not welcome him into the lodge. Yet before any confrontation could begin, he too fell silent, listening along with his shaken companion to the sounds drifting in through the walls from the outside: a quick snick of compressing grass, the burble of gravel dislodged under shallow running water at the stream crossing—footfalls. Several animals—three, or four. Quick, short strides—not deer. Sure steps, heavy—not coyotes. The sound of noses filtering scents close to the ground, a curt snarl as one creature responded to a nip from another—wolves.

Two very cautious beavers emerged late that evening to a silent landscape. The silence was to continue as winter extended into spring. The hooved hordes had all but disappeared; there was no place for them to hide in the denuded woods. Individual grazers rarely showed themselves, pausing only briefly with heads down in search of food while watching their surroundings through the corners of their eyes, finally spooking each other into flight.

As the days warmed, a low skein of green spread over the ground. Willow shoots emerged near the shore and bent toward the zenith. The beaver found occasional forbs pushing up beyond the waterline—a food he preferred to last fall's bark stores. He was freed to use his cache of sunken branches to patch the dam. But he took no risks, so sure was he that the woods had been invaded by wolves.

Since their arrival, he had twice caught the taste of blood in the water. And though with his limited sight he never saw them, he listened carefully, late on the nights of the full spring moon, for sounds infiltrating the woods. Behind the distant surf of the main stream through the valley bottom—now swollen to a river rumbling at flood stage—he no longer heard the disharmonious yips and yodels of the coyotes. Their choruses had been replaced by the mournful, chilling call of the wild.

His pond mate disappeared. Perhaps he had moved on. So the beaver worked alone in silence, always monitoring the brace of changes in the forest—new scents, altered sound quality as the plant cover advanced, new flavors in the water. He found rushes and violets and tree seedlings in this year's bloom of spring foliage, bounty that he had never seen when every day began with deer or moose or elk on his shore. His dam grew sturdier—the waters rising behind it as the spring runoff strengthened. The pond margin advanced, giving him access to shoots of birch and dogwood that grew quickly in the absence of the grazers that once took their leaves.

One afternoon in the heart of springtime, he found another beaver crawling over his dam. Conditions in the valley had improved enough that he would no longer have to tolerate another in his domain—this interloper could move on and build another dam. Yet the newcomer did not hurry on through the pond like other passersby but tarried, sizing up the waterworks. The new beaver barely seemed to notice him, reacting to his aggressive approach as if it were a bluff soon to be dropped. And soon he did drop it, as he found that this young beaver was a potential mate. He followed behind as she inspected his wide patch in the stream. It had grown substantial enough to share.

As his pond deepened, the waters reflected changes in the fertility of the valley all around. He and his new mate saw tadpoles for the first time, and a muskrat hole appeared opposite the lodge by the stream inflow. Striders and whirligig beetles dimpled the ripples as he swam past. The crowns of the trees filled out, partially shading the waters behind the dam—their branches alive with unaccustomed birdsong.

One bright morning early in the summer, he returned to the lodge to find his mate hissing at him and barring his entrance to the high dry chamber. He did not rise to the challenge, but backed out and eventually spent the next weeks living in the hole in the bank. When he was finally welcomed back into his own home again, he found four new kits who would now be sharing the den with him and his mate.

Over the rest of the summer they worked together. The kits enlarged the inner chamber of the lodge, while the two adults enlarged the fortress from the outside and raised the dam. The impounded water level rose, driving the shore farther back, increasing the measure of security from the dangers in the woods. As fall approached, the beavers laid in extra provisions, sinking a pile of leafy branches into cold storage in the deepest center of the pond.

The winter would be less secure for the wolves. The grazer population began to falter while the wolves were supporting prosperous litters of pups, raising their own population toward its maximum. As the days grew short, the wolves began to redirect the strain of their occasional hunger toward each other in extended dominance disputes.

When the thaw finally began, the beavers emerged to a bright new season. The threat of ambush kept them constantly on guard as they foraged on shore, but there was a benefit to the advent of the wolves. The predators had brought a heightened spring to a forest that the grazers had before kept in a perpetual state of leafless winter. Spicebush, blackberries, and horsetails filled in the gaps, providing cover between new saplings. Crayfish appeared amid the sunken footings of the lodge. On warm nights, the beaver stuck his head out of his pond to find a treble chorus of song clinging to the surface—his ears rang with the omnidirectional throng of spring frogs drowning out all other sounds.

Swimming behind the dam, he was momentarily disoriented by the effervescence of a hatch of mayflies coming up all around him. Tens of thousands of them had prospered in the enlarged pond—growing unnoticed on the bottom since last year, and now their synchronous emergence had begun. They paused on the surface to spread delicate wings, then lifted off to continue spiraling upward into the air, soon to return and put millions of eggs back into the water.

The forest thickened, as did the overlapping calls and stridulations arising from within. The beavers grew familiar with wood ducks leading rafts of chicks across their pond, now shaded by a canopy of green. The flow in the water was stronger, colder, steadier, with fewer sudden drops or crests—the stability resulting from other new waterworks that were taming the waterway throughout the valley.

The beavers bore another litter—three more kits. An industrious family of nine now raised the lodge level and extended the dam past landmarks that had stood high and dry above the previous bank. Now the streamside trail and a reach of the forested margin beyond were on the pond bottom.

A shallow, sunlit meadow behind the old bank was swamped by rising water, still only a few inches deep—creating prime habitat for willows, a favorite beaver food.

In his expanding wetlands the beaver encountered thousands of mosquito wrigglers. For months their numbers increased, the adults constantly pricking at his nose and swarming around his eyes, their whine imposing in his ears. At dusk and dawn, they were persistent enough to find their way through several feet of interwoven branches into the beaver den inside the lodge. Where they were thickest, the only escape was to slip underwater.

The mosquito population continued to swell, until the bats came back. The beaver could hear bats inches above the water cutting across his path as he swam across the pond at night. After they returned, the wriggler populations began to decline.

The wolves had facilitated the prosperity of the beavers, but the beavers would slight their benefactors. The mosquitoes that bred in the swampy places behind the enlarging dams up and down the valley now carried a microscopic fluke parasite. It had entered the mosquito population through a single mosquito feeding on a defenseless coyote, mauled by the wolves, dying in shelter beside a log.

Coyotes spent little time by the stream, preferring windy, open land. But the wolves removed the last of them from the forest margin, allowing the foxes to occupy a broader range and increase their own numbers. Foxes lived in the still, humid air under cover and were easier for the mosquitoes to find. And so the fluke parasite, which had little effect on the mosquitoes, was transmitted first to a pair of foxes, then picked up again from their blood by other mosquitoes who spread the flukes to the wolves. The flukes swam through the veins of canine hosts and eventually colonized the central regions of their circulatory systems. They caused little harm in the foxes, but the infection gave the wolves heartworm disease.

That was not the only such problem the wolves had. One of them had acquired another malady by ingesting cysts from the lung tissue of a parasitized elk. The encysted parasites survived the digestive process in the wolf stomach, then hatched in the intestines. This parasite was nearly asymptomatic in the grazers, but in the wolf, it caused a debilitating intestinal tapeworm infection. The infected wolf continually shed thousands of tapeworm eggs in its scat. As the season progressed, the deer chanced to graze up a few such eggs, which gave rise to populations of the parasites encysted in their lungs, which, in turn, led to the spread of the infection throughout the rest of the wolf population in the valley. With prey more and more difficult to find, the additive stresses set the scene for a wolf population crash. The only question was, What form would it take?

The third spring after the wolves came, the kits from the first beaver litter were ready to strike out on their own. When the creek's flow was at its highest, they headed off upstream. The flow was strong throughout the valley, and as the tide surged freely across the spillways, scores of muscular trout leapt headlong against the overflow, over one dam after another. They darted through the channels, snapping at drifting targets before vanishing upstream following the departed second-year beavers.

The night air no longer thrummed with the distant surf of a river at spring flood stage. Beavers had engineered dams up and down across the valley floor to tame and contain the cataract. The new chain of impoundments continually recharged the groundwater, greening the forest. As the dams grew taller, the water depth had a chilling effect on the deeper currents behind them. The cold drove out warm-water species of bottom-feeding fish that had encroached from the flatlands far below.

Sediments fell out of the water in the stilling ponds behind each dam, leaving the river clear and pure. The flattest parts of the oldest ponds filled in with silt, shouldering the streambed laterally across the valley floor, leaving raised terraces. There, the willows gave way to aspen. The beavers took advantage of this new white-barked resource by digging canals across the flats, through which they floated building materials and food stores back to their ponds.

The tremendous cloudbursts over the headwaters far upstream in late summer no longer resulted in torrential runoff deepening a narrow streambed. The water now rose gradually to new high levels in each successive pond. The serial waterworks let the valley keep the extra rain, absorbing it into the water table. So the trees grew their roots deeper, and their crowns taller, and all that depended on them prospered in turn. The rocky, steep pinewoods center of the valley was gradually transformed into terraced, deciduous shady bottomlands.

On a brisk day late in the fall, the beaver came upon a scene he could not comprehend. The grass was flattened, the ground stained red, and barely recognizable parts of a dismembered wolf were scattered here and there. The beaver sat motionless, his nose working through the visceral scents, his ears on guard for the slightest sounds. Then, without conscious command, the growing fear began to move his feet back toward the safety of the water.

The wolf population had risen to the level at which fights over dominance and territory were common. At the same time, the wolves were debilitated by disease and forced by the crashing grazer populations to move farther and farther afield on hunting forays. These stresses came to a head that winter, and the wolves left the valley. The beavers did not notice, since they were frozen in, living off the bark in their sunken cache of branches stored under the ice—their lodges heated by the warmth of their own bodies. Their first clue that things had changed came with the appearance of a few deer the following spring, grazing unconcerned at the edge of the pond.

Those returning grazers found the abundance of an Eden in the valley. They responded to the green bounty by reproducing as fast as they could. In a few seasons, their numbers were out of balance with the carrying capacity of the land. Even as the signs of overcrowding and competition began to show over the years, they continued to increase their population as if they would again soon find their lost days of plenty. When the inevitable came to pass—and the cycle had swung—scrawny grazers would be reduced to eating bark once again. What the grazers would finally reap from their profligate ways would be wolves.

For their part, the beavers imposed a stabilizing influence on the valley. Their population would grow too but eventually crest and be self-regulating thereafter. They maintained a single family per dam, with no dam in sight of another. Their own food prospered in the habitat they provided for it: cattails and water lilies, and thickets of willow and aspen, alder and birch. They transformed the landscape far upstream to the rocky limits of the headwaters cataracts and downstream to the lowlands plains.

The beaver and the wolf were keystone species in these woods. The consequences of their actions rippled out into the forest in disproportion to their relatively small numbers. In an environment managed by the builders of the dams, the woodland residents eventually came into balance: porcupines and flying squirrels, warblers and bats, weasels and bobcats, as well as the larger grazers and their predators. Fish colonized the streams, bats filled the air, and together they moderated the numbers of mosquitoes. Life continued to be a struggle for the individuals of every kind, as it always has been and will be. But the crests and valleys in the yearly population numbers would even out, providing no numerical advantage to any one species but maximizing the sum of all.

Science Notes

"I have watched the face of many a newly wolfless mountain, and seen the south-facing slopes wrinkle with a maze of new deer trails. I have seen every edible bush and seedling browsed, first to anemic desuetude, then to death. I have seen every edible tree defoliated to the height of a saddlehorn" (Leopold, 1948: 139). These words describe the extreme cyclic swings that reveal the influence of the wolf upon everything in its habitat. The creatures in the habitat make up a trophic cascade, a food web across which changes in the status of one key member can affect other members several links away, members with which the key member is not in direct contact. For example, the number of green leaves in spring is directly proportional to the number of wolves in the forest. This correlation is due to the link between herbivore damage and the number of predators of those herbivores (Ripple and Beschta, 2006). The dynamics of food webs that include wolves have been studied on Isle Royale in Lake Superior (Peterson et al., 1998), as well as in Yellowstone National Park (Phillips and Smith, 1996). The apex predator is a keystone species—its effects ripple across the biological landscape. Another keystone species in this habitat is the beaver (Naiman et al., 1986). The wolf and the beaver impact trophic webs from, respectively, top down and bottom up. Beavers erect dams that impound settling ponds, which accumulate stream-borne silts, thus clarifying the water. Their continuous efforts leave the valleys layered in rich, well-watered soils— maximizing the diversity and productivity of the flora upon which the rest of the bottomlands food pyramid stands.

References

Leopold, A. S. 1948. *A Sand County almanac*. New York: Oxford University Press.

Naiman, R. J., et al. 1986. Ecosystem alteration of boreal forest streams by beaver (*Castor canadensis*). *Ecology* 67:1254–69.

Peterson, R. O., et al. 1998. Population limitations and the wolves of Isle Royale. *Journal of Mammology* 79:828–41.

Phillips, M. K., and D. W. Smith. 1996. *The wolves of Yellowstone*. Stillwater, Minn.: Voyageur Press.

Ripple, W., and R. Beschta. 2006. Linking cougar declines, trophic cascades, and catastrophic regime shift in Zion National Park. *Biological Conservation* 133:397–408.

Where Nothing Grows

In the midst of the humid low forest of eastern North America, gaps open out among the trees—places where, in contrast to the rank foliage all around, nothing grows. At the edges of these gaps, grasses and forbs compete for the space, backing up into the trees. But sunlight falling on the bare patches of wet soil in the open energizes no green leaves—the ground there is sterile.

Under the surface the earth is boggy. Mats of plant matter have accumulated over the centuries and decayed in place, souring the soil until the decay process itself is slowed by the acidity. In this groundwater acid bath, the skeleton of a fallen animal—including the teeth—would dissolve without a trace in less than a season. Minerals essential for plant growth have been leached from the soil. Seeds that fall in these open spaces fail to germinate.

These bare patches are a permanent aspect of flat, well-watered coastal lowlands. But the uncontested growing space in these sun gaps does not go unused. A few types of plants can survive in the nonnutritive soil. Most successful among them are the pitcher plants. The pitchers have developed a way to compensate for the lack of essential elements in the soil—they leach nutrients from the bodies of insects they have captured in the pools within their flagon-shaped leaves.

Pitcher plants grow too slowly to compete with the tangle of greenery at the edge of these open spaces. But out on the marshy soil, where potential competitor plants die for lack of nutrients, they prosper and diversify. As the sun reemerges after heavy rains, the surfaces of their water pockets glint across the boggy depressions. Should a beetle slip off the treacherous lip above one of those little cisterns, or a water bug land on the reflective surface, it would immediately fall through the waterline. Its slowing, futile movements would fade into the depths of the digestive pool, eventually providing the pitcher with the sustenance to grow where little else does.

Every pitcher plant is a self-sustaining microcosm. The little basin is home to a food chain based on bacteria and extending up to top predators, the larval stages of specific, coadapted midges and mosquitoes. The plant provides the trap that captures the prey, and the animals within each play a

One false step. A cricket explores
the edge of a purple *Sarracenia* pitcher.

part in reducing the bodies of those insects all the way down to the soluble nutrients that support the plant in which they live. To prevent the pool from becoming stagnant, the pitcher plant absorbs carbon dioxide from the water and infuses oxygen.

A very different environment thousands of miles away shares one characteristic with the Eastern Seaboard—it contains a habitat where nothing grows save for a pitcher plant. This forested domain falls along the mountains on the border between northern California and Oregon. It is a sloping, rocky terrain of musical brooks, where bracing breezes sweep up out of nowhere in the middle of a calm day. Alpine meadows open out onto hundreds of square miles of dark conifers that carpet the slopes below the tree line, shot through with yellow veins of aspen as the brief summer wanes.

This realm grows as much biomass per acre of woodland and as much floral diversity per foot of trail as would be found in any tropical setting, reflecting the confluence of its three habitat zones—the granitic Sierra Nevada to the southeast, the metamorphic Coast Ranges to the west, and the volcanic Cascades to the north. Magnificent firs and cedars sigh in downdrafts flowing from the peaks; ospreys and pine martins keep watch through soaring branches; mariposa lilies and tiger lilies rise through carpets of fallen needles, sharing the understory with cobra lilies, which are not really lilies at all.

Cobra lilies are slender pitcher plants whose hollow stalks are topped by a bulbous helmet that conceals the water column within. In profile, a cobra lily resembles an angry cobra reared up to display its inflated hood. An entrance through the underside of the helmet meets insects that walk up along a paired, leafy appendage reminiscent of a serpent's forked tongue. The stalks of cobra lilies twist as they rise, so that clusters of plants with a common footing each face a different quadrant of their surroundings—like a family of vipers ready to strike in any direction.

Insects on the sweetened, forked tongue of the cobra lily look up at the inside of the hollow helmet as they sip the scented nectar at its opening. The upper surface of this helmet is patterned with translucent patches like rows of skylights, where the green matter has been resorbed. Flying insects respond to brightness above, expecting that the breeze waits in that direction—if they fly upward, they expect soon to be borne along on the wind, following a levitated scent trail to their next meal or mate. So, when it is time to fly, they usually rise.

The bright roof of the cobra dome is in exactly the opposite direction from the darkened portal below through which the insects gained entry. They have never experienced this conundrum, and, like lobsters in a lobster trap, they do not have the capacity to resolve it. Only by chance would they stumble on the exit. More likely, they will spend the morning exploring the ceiling and eventually wander down the well-lit, increasingly precipitous walls of the tubular stem, soon losing their footing among the thin, ever steeper, downward-pointing spines. Then they will fall into a chasm too narrow to allow them the headspace needed to fully extend their wings and gain lift. They will finally splash down against a pile of insect parts coated with a deadly solution too slippery to escape from before it permanently stifles their efforts.

Cobra lilies contain surfactants in the water that fills the base of their hollow stems. Less than one part per thousand of these wetting agents will change the characteristics of water so that it clings to, rather than beads on, an insect's waxy cuticle. Insects that usually bounce off water instead fall through the liquid surface.

An insect drowns in microliters of water, enough only to wet its skin and cover its spiracles with a thin film. The surfactant-laced water in the base of the cobra lily rises up around and embraces an insect, stilling the victim in seconds if it merely touches the oiled surface.

A cross-section through the base of the hollow stem of a cobra lily exposes a compacted plug of the leached hard parts of scores of flying insects. Barely recognizable are the remains of midges and solitary bees, shield bugs, soil flies, weevils, gnats, and day-flying moths, all compressed into a tangled mat of interwoven legs, head capsules, and wing covers. The bristling thatch is blackened by the digestive work of bacteria that live in the depths of the pitcher and solubilize the insect protein for their host plant. Mere micrograms of nutrients are absorbed as a result of each brief moment of capture. But that is more than most roots could wrest from the unique soil beneath a glade of cobra lilies. That soil stills the growth of other plants, allowing the slower-growing pitchers to flourish.

The thick curtains of conifer forest covering this Northwest temperate rain forest are occasionally parted by sun gaps opening above rocky meadows where very little grows. The rocks in those meadows are magnesium silicate deposits of serpentine.

Serpentine is not produced in the mountains. It forms under incomprehensible temperatures and pressures in volcanic vents deep beneath the surface of the ocean. The hydrothermal forces that create serpentine are so great that, at the time of its creation, quartz runs through its veins as a liquid. The rock's waxy surface, variegated in darkening jade shades, is often crosshatched with a thin lacework of milky white crystalline quartz—frozen in place at the instant of its deepest penetration.

After it cools, serpentine is transported toward land by the process of seafloor spreading—riding for millions of years upon moving tectonic plates. Eventually oceanic serpentine is rippled into thin bands as the rocks of the seafloor crash in slow motion onto the edge of the North American plate. The tectonic energy of this collision thrusts the rocks above the water, generating the coastal range mountains.

Smooth, polished green outcrops of serpentine come to lie exposed thousands of feet above sea level, high on forested slopes—the salt waters of their past replaced by sweet snowmelt. High concentrations of magnesium leach from these bands of fossil seafloor. The mineralized water that flows through serpentine deposits interferes with the uptake by plant roots of an essential element, calcium. As a result, serpentine soils are barren of the growth that flourishes just beyond their margins.

Snowmelt percolating down from mountain ridges far above tree line emerges on the lower slopes as cold alpine springs. When these waters have infiltrated serpentine during their subterranean descent, the streams and wet meadows they recharge are not crowded with stands of willow or alder, not thick with gooseberry or mountain ash, not shaded by any serpentine-intolerant growth. The banks of such mineralized glades are perfect cobra lily habitat. Green helmets rise by the thousands here, standing shoulder to shoulder where gaps in the tree cover let in the light—their roots bathed in algae-free, crystal-clear running brooks.

Plants that depend on soil nutrients could not survive in these glades because of the chemistry of the water, but these pitchers get their sustenance from the insects they capture. The ranks of cobra lilies flourish beneath the waving metronomes of their own bulbous flowers, which nod on stems that hold them far above the crowd of carnivorous pitchers. The stems are two or three times longer than the plants' own stalks—which prevents the pitchers from eating their own pollinators.

Cobra lilies prosper in this niche in isolated colonies throughout the Siskiyou and Klamath Mountains of southwestern Oregon and northwestern California, northern California's Trinity Alps, and down into the California Sierras. They also occur closer to sea level in the Rogue River drainage of Oregon farther to the north, where they inhabit a bog niche like the one occupied by their marshland cousins far to the southeast.

The native pitcher plants across the continent on the Atlantic seaboard have struggled to survive in recent times. Their habitat is contracting—less than 3 percent of their original range remains. Their marshes have been drained and replanted in monoculture slash-pine tree farms or paved over with layers of asphalt by commercial expansion. But the cobra lilies on remote mountainsides far to the west rest under deep snows over a long winter. Their rugged habitat is inaccessible much of the year to vehicle or even foot traffic, insulating them from the pressures of civilization. These habitats have been further protected by their designation as state or national wilderness preserves. Designated wildlands are prevalent in the western United States, ensuring the survival of diverse communities—of plants, and the animals that feed off them, and vice versa.

Science Notes

Plants have evolved strategies for occupying almost all the sunny spots on the Earth. Insectivorous plants, such as the northern pitcher (*Sarracenia purpurea*) and the cobra lily (*Darlingtonia california*), have occupied spaces where the groundwater is, respectively, too acidic or too high in soluble magnesium to allow the roots of most other plants to function (McPherson, 2006). Though many pitcher plant species produce their own digestive enzymes, these two rely heavily on a mutualistic relationship with a guild of microbes and arthropods adapted to live in the pitcher and digest the prey that falls in (Heard, 1994). This inquiline community includes bacteria, protozoa, rotifers, and the larvae of several diptera species. These larvae occupy distinct niches within the pool, either feeding on the sunken detritus of older prey or the suspended bodies of recently captured prey, or filter feeding on the lower members of the indigenous food chain (Heard, 1994). The pitcher regulates the oxygen and carbon dioxide levels, preventing the water from stagnating (Bradshaw and Creelman, 1984). Pitcher plants can meet 70 percent of their nitrogen requirements from the prey they capture (Schulze et al., 1997).

S. purpurea is endangered across much of its range in the Southern and Midwestern United States, as are other *Sarracenia* species. In the Northeast, the plant has reacted to the infusion of nitrogenous pollutants from the air by ceasing to produce pitchers, and producing flat leaves instead (Ellison and Gotelli, 2002). This removes the niche in which the pitcher inquiline communities lived. The Wilderness Act of 1964 formalized wilderness designations of U.S. lands. There are 16.5 million acres of such protected wilderness in California and Oregon—eight times the area protected in the combined Atlantic seaboard states. *D. california* was removed from the endangered list in 2000.

References

Bradshaw, W. E., and R. A. Creelman. 1984. Mutualism between the carnivorous purple pitcher plant and its inhabitants. *American Midland Naturalist* 112:294–304.

Ellison, A. M., and N. J. Gotelli. 2002. Nitrogen availability alters the expression of carnivory in the northern pitcher plant, *Sarracenia purpurea*. *Proceedings of the National Academy of Sciences, U.S.A.* 99:4409–12.

Heard, S. B. 1994. Pitcher plant midges and mosquitoes: A processing chain commensalism. *Ecology* 75:1647–60.

McPherson, S. 2006. *Pitcher plants of the Americas*. Granville, Ohio: McDonald and Woodward.

Schulze, W., et al. 1997. The nitrogen supply from soils and insects driving growth of the pitcher plants: *Nepenthes mirabilis, Cephalotus follicularis*, and *Darlingtonia california*. *Oecologia* 112:464–71.

Part 2. Air

The big picture. A dragonfly pauses
to take in the (nearly 360-degree) view.

Eye of the Needle

The streamside trail flattens. The water quiets and opens onto a still, shallow pond. You leave the ferns and the salamanders to their deep shade and find a seat on a log among the reeds and sunflowers by the shore. If you are quiet, the calm will return and you will be accepted into the scene. Finally, you extend your finger, as if pointing at the far side, and you wait.

And if you are patient enough, a dragonfly will alight right on your fingertip. Earlier in the season, it may be a powder blue or a flame orange skimmer—harbingers of the springtide. In the height of summer, perhaps it will be a big, aggressive green or blue darning needle.

And if you have an unusually steady hand, you can start to ever so carefully draw your visitor in closer. Hold your breath and pull it right up face-to-face, and you will be rewarded with the rare chance to gaze squarely into eyes as hypnotic as a revolving mirror ball—but even better, for these entrancing spectacles of biological engineering are alive, and are looking back at you.

The dragonfly's eyes cover most of its head with a spherical latticework of thousands of crystal facets. As you gently rotate your finger, you will see patches of color that stay centered in each of its eyes, no matter which way it turns. Though these eyes have no moving parts, two spots—one a bright highlight, the other a thin, deep shadow—fixedly hold the centers of their revolving orbs. These two marks travel across the compound eye, moving against the direction of their rotation, and—from your perspective—staying centered.

One of these spots, a bright glint that holds its place above the center point of each eye, is an image of the sun. It stays fixed at a constant angle of reflection. The other, darker patch that stays positioned just below the sun glint arises from within—it is an image of the dragonfly's retina. It is as dark as a pupil—light that falls into that spot does not return but is absorbed and processed by the animal's brain.

These two points—one of light, one of shadow—are constant features of the surface of the compound lens. They don't move as the dragonfly's head moves, giving the eyes the illusion of depth. It's as though you were watching spots that moved more slowly than the surface because they were not on the surface, but deeper within. Yet the textured globe of those eyes is opaque, and pigmented in the colors of the animal's flanks and wings. You cannot see through it.

The dark pupil is a fiber-optic projection onto the finely patterned surface. It is carried in transparent strands connected to the brain below. Those optical fibers radiate out from their center in all directions like the spokes of a dandelion seed head. The dark window through them is visible only on the few lens elements that are pointed directly toward you. The deep image appears, then disappears from each fiber tip as it rotates in, then out of alignment with your line of sight.

The shape of this dark pupil evolves as your viewing angle changes. The light-carrying fibers have been arrayed to optimize the insect's command of its air space. So as you gradually rotate the dragonfly to face you, aligning your line of sight with its line of flight, the dark pupil grows larger—like the eye of a predator dilating when prey comes into view. Rotate the creature further, and the dark region contracts to a thin cat's eye that glances to the side and behind. Track the dark image of the retina up the mosaic lens to the top of the head, and the geometry of the matrix tightens until the pupil contracts into a single black point; from that angle, the insect is squinting at the noonday sun.

These biological star sapphires take longer to mature than do the eyes of most insects. Though dragonflies live only briefly as adults, their lifespan is quite long by insect standards. Season after season during the ice-free days of spring and summer, dragonfly nymphs develop underwater, stalking the streambed. The nymph may grow for up to five years before it emerges through the surface to begin its week or two of adulthood. It will have snared hundreds of mosquito larvae, mosquito fish, and other prey, and will have survived a gauntlet of larger predators swimming above it on the food chain. Only rarely does one of the eggs laid by the last generation survive through all those summers, finally spreading its cellophane wings to taste the air.

Looking out through those formidable eyes, the dragonfly commands a worldview we would find daunting indeed. Our brains use a different system of perspective, and we could not cope with the flood of visual information that these stream skimmers receive every moment. Dragonflies do not see the world divided into thousands of facets; their brains integrate the input from all their lenses into a seamless three-hundred-degree moving panorama. They view the day through a fish-eye lens, experiencing every direction simultaneously, constantly, continually.

They live in a vista dome stereoscopic in its depth across the midline of their sight, an IMAX presentation both front and back. On the wing, they watch the shoreline willows and horsetails approaching, watch them accelerate to glide past when closest, and watch their profiles fall away behind—all at the same time. The scrolling tableau is anchored by twinned images of the sun, one implacably fixed against the blue sky above, the other a reflection skidding past on the surface of the water below.

The reflected sun holds position on a transparent mirror of water rippled by crayfish, dimpled by water striders, and striped with cattails broken by diffraction where they pass through the surface. Dragonflies are the most active and agile of the airborne insects. As they bank, accelerate, and switch back and forth across the sky, the surrounding riverscape spins and twirls, tumbling from one course to the next.

The dragonfly's head swivels while the insect rests on your fingertip. The creature has noticed a mote in a shaft of sun slanting behind the rushes. It can spot an aphid rising three meters away well enough to distinguish the edible from the distasteful species. Suddenly—with a barely perceptible touch—it lifts off, instantly flying at full speed, and is away in pursuit of something only it sees.

You cannot experience the vitality of the dragonfly's eyes by studying photographs. Photographs are two dimensional, and the refracting highlights, the illusory depth of these living jewels, cannot be gauged in a motionless flat depiction. Neither can their form and function be grasped by the study of a mounted specimen, for the eyes opacify soon after the creature dies. Yet the chance to fully appreciate those eyes is well within your reach; it is waiting for you by the side of a stream—right at your fingertips.

Science Notes

Arthropods have evolved eyes that work quite differently from the focusable lense eyes of the mollusks and vertebrates. The compound architecture (Stravenger and Hardie, 1989) of insect eyes is easily seen in dragonflies, which have the biggest such eyes of all the terrestrial arthropods. Those eyes can have up to thirty thousand facets. What feelings would have been inspired if we could have made eye contact with *Meganeura monyi*, the dragonfly from the Carboniferous period that was the biggest insect ever known, with a wing span of two feet?

References

Stravenger, D. G., and R. C. Hardie. 1989. *Facets of vision*. Berlin: Springer.

Spider on the Fly

Though spiders are walkers, nimble afoot and wingless, they are not shy about taking to the sky. They spend much of their time suspended weightless and are always ready to take off. They may patrol gravity bound across acres of trackless foliage, but if their course leads them to a precipice so steep their tiny eyes cannot see the bottom, they step over the edge and off into space without breaking stride. As quick as a spider step, they affix a lifeline at their feet, jump as far as they can, and sail through the air with eight legs spread wide. Then they pull their tether tight to arrest their fall, turning their trajectory into a broad, swinging arc. Spiders fly—on diaphanous wings of spider web.

They spin that web constantly—fine protein strands recycled from their high-protein diet. The line functions as a ninth appendage upon which they are utterly dependent. Even on solid ground, spiders prefer to stand on quivering strands of silk. The most terrestrial wolf spider spins a few strands around his bivouac and places a forefoot on one of them to serve as his ears. Telltale vibrations transmitted down the line tell of wind and weather conditions, signal the passage of potential mates or prey, and warn of the approach of danger.

A spider makes minimal weights of web at a time—mere micrograms—yet she can play it out yard after yard. At daybreak, she collapses all the web in an orb that covers a square meter, gathers it into a ball no bigger than she is, and then eats it all in a few minutes (unless it has captured too much dust). The following evening she recycles her work, extending a very small mass of web very far by spinning it very thin. The web is but a fraction of a micron thick, too slender for prey to see (or easily avoid).

We see the invisible threads only from the way they bend—diffract—light. At low sun angles, a meadow's contours come alight, outlined in scores of glistening filaments—spectral wrapping that traces spider paths across the land. The silver strands rappel diagonally through space from branch to branch to blade of grass. A hundred spiders could hunt each square meter of meadow, and all of them would escape notice but for their shimmering draglines.

Spider web is the quintessential expression of the forces that hold the molecules of life together. These life forces follow an inverse distance law: they are stronger where the distances they span are shorter, like the force in the surface tension of water. The power of molecular bonds is relatively immense over extremely short spaces, such as the distance between two surfaces in touch with each other. A small moth accidentally flattened onto the surface of a streamside puddle will be held there forever by a power much greater than the insect's strength could ever overcome.

Bonds like those between water molecules hold the proteins of spider silk together in their polymers. They provide the adjustable links through which the elastic protein units instantaneously re-form when they slide past each other. And they provide the force that holds the web to whatever it touches.

The momentum of a hurtling insect sends ripples out across a transparent orb of spider web from the point of impact—ripples that echo back and forth through the network, gradually fading until all motion ceases. The insect has just splashed into a virtual vertical sheet of clear water and is held to each strand just as firmly as it would be by the surface tension of a streamside pond.

But the web lacks the total surface area of a pond, and an entangled insect works the hundreds of microscopic bristles on each leg against the strands to free itself. The spider must find and secure its prey before it escapes. Finding that prey is complicated by the geometry of the web—the spiral structure distributes the vibrations of a struggling insect uniformly across the disk, just as it distributed the energy of impact from one strand to all.

Nonetheless, the spider at the center can interpret the omnidirectional signal. She has an appreciation of vibrational harmonics as keen as that of a maestro violinist. She will not be drawn to low-frequency vibrations—she ignores the snagged feather twisting in the wind or the ant testing a holdfast point. But signals in the hundreds of hertz—the buzzing of struggling prey—spin her around the center of the web like a compass needle orienting to north. Plucking the radial strands, she measures the dampening of the return wave—an indication of weight on the line—to find the direction toward which to pounce.

At the tip of a tall branch sits the spider. Her weight bends the outermost leaf downward into a gap between trees that reach toward each other across a woodland stream. The canyon of green before her is too broad for her little eyes to see the far side. Nonetheless, she feels the afternoon

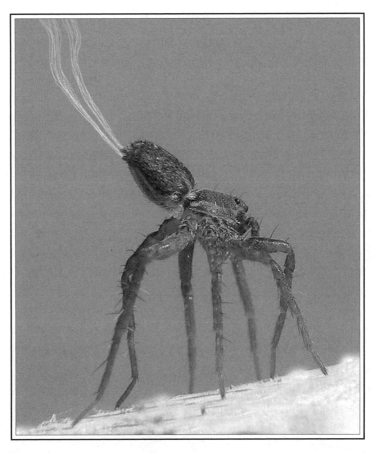

Ballooning.
A wolf spider prepares to test the wind.

breeze flowing upstream through the gap, much stronger than in the dense foliage behind her; the air current is enough to float a parade of small, wind-borne insects through the gap. Mass-produced midges and aphids, soil flies and mayflies that balance their high attrition with their even higher numbers are her primary prey. She will cast her net there, after she manages to span this promising passageway.

Holding her abdomen high, she shoots a strand of fine, sticky web straight up, sending a streamer out on the wind. It floats away, its end snapping back and forth, elongating over the water as she continues to spin. After less than a minute, she stops and walks away upwind until she feels resistance from the extended line—it is pulling against her, stuck to something solid.

She anchors her end, along with the end of a new, nonsticky line, to the leaves. Then, spinning more as she goes, she hangs her weight on the sticky strand and carefully climbs out on it through space. Pulling out stronger cord while she clings to the anchored high wire, she displays her spider talent of walking on the sticky strand and not getting stuck.

At the far side she attaches a second guideline to the reinforced strand and heads back, spinning a loose loop out below her. Affixing the near end of the dangling loop, she slides back down it, her pendulous weight pulling the looser thread into a *V* below the top line. Where she comes to rest at the low point of the *V*, she attaches a third strand, and, supporting herself by her hind legs and her spinnerets, she drops headfirst into space on an elongating filament.

Halfway to the water she arrests her descent and hangs in midair from her slender tendril, waiting for the breeze to push her away from the vertical—sailing her out, swinging her back. After a few minutes, a random gust slaps her into something hard, which she grabs with two, then six feet (the other two holding her lifeline). There she affixes her third guywire, completing the central frame of a bridging orb that will soon span this gap above the stream.

In a few more minutes, she has emplaced the rest of the spokes. Then, starting at the center, she adds widening arcs of sticky silk, made of the most tenuous of threads. Any single one of these strands would be useless against the passage of a large, onrushing beetle or hornet—snapped in two. But the spider designs the structure so that all the strands will pull together, distributing the force of an impact on any one strand to all. Drawing on the millions of years of experience inherent in her species, she builds a net whose resistance is perfectly balanced by its flexibility. All strands will work conjointly to stop the fastest and most massive prey, the finished wheel a spiral work of engineering genius.

When a spider feels the need to go for a sky ride, the habit of casting a streamer on the wind takes on another dimension. Spiders sit low on the food chain and rely on flight, rather than their bite, to save themselves. At the appearance of a threat (often in the form of a bigger arachnid), a small spider backed out onto the edge of a cliff pirouettes on the ledge to face the wind, raises his abdomen high, and in seconds broadcasts a bloom of silken lace onto the breeze. Then he leaps untethered into space and floats away, coming up suspended from the gossamer parachute he has just broadcast, hung by his silken line. He climbs to the center of the flimsy raft and continues to spin, fabricating a tenuous airfoil that will catch the wind and free him from the ground, carrying him he knows not where.

Spiders are the best of fliers. They and the web they spin present the wind with the greatest surface-area-to-weight ratio of any animal that takes to the air. As wind riders, spiders are completely passive—no energy is required for them to coast on the breeze, so their airborne endurance is transcontinental.

But when spiders vacate an area in search of more promising hunting grounds, they travel with more direction than the passive parachute of the dandelion. Migrating spiders optimize their success by choosing for their departure the best hour (when the air is warm and the currents rising), the best place (a promontory), and the best instant of launch (when a passing zephyr gusts past from below).

A floating spider standing among the threads of his woven sky boat balances on the shifting matrix. He responds just as an orb spider does in the center of a spiral wheel of silk buffeted by the breezes that rock the branches to which it is attached. He paces, like a captain measuring the foredeck of his galleon, and inspects the rigging. He can eat the knots if the web tangles and can spin fresh, fluffy sheets as he glides along below the clouds. And he captures any tiny flying insect that may blunder into his webbing. He has a free pass on the will o' the wind, cruising his traveling driftnet through the currents of sky plankton. But if the zephyr he rides sweeps him up tens of thousands of feet into the stratosphere, decorating his frozen body with a starburst of dry ice crystals, the journey is over. Spiders can stand the cold only when insulated beneath the bark or the ground—they cannot revive once they freeze solid.

Charles Darwin witnessed the invasion of his sailing ship *HMS Beagle* by a flight of spiders two hundred miles off the coast of South America. Such mass migrations demonstrate the potential of flying spiders. They are the first colonists to arrive on remote islands emerging from the ocean, or on barren fields of freshly cooled volcanic lava. Rising ever higher through the sunrise, settling to lower elevations by evening, the tiny fliers drift in loose formation a mile above the sea, their lacy, air-built balloons catching the amber light. They are watching from above, waiting for a hole in the wind through which they might drop in on promising new hunting grounds—sites from which, perhaps, they will not soon have to fly.

Science Notes

Spider web has great linear strength (Craig, 2003) derived from the molecular surface tensions of its water molecules, bound to each other at the water:air interface. But strength comparisons of spider web with steel cannot be meaningfully scaled up to cable dimensions; at that width the contribution of the surface tension of the water molecules becomes negligible.

References

Craig, C. 2003. Spider webs and silks: Tracing evolution from molecules to genes to phenotypes. New York: Oxford University Press.

Sky Walkers

The margay walked just above the ground, on a low branch so thin it quavered with her every step. Her legs flexed and her long tail switched from side to side to counter the swaying; she compensated so well that her head glided as steadily along as if she were on solid ground. She was an acrobat, lithe and sure, as accomplished a hunter in the trees as on land. She could hang from one back foot like a monkey as she negotiated the scaffolds of her world. When she paused to scan the streamside just below, the pattern of leopard rings on her coat merged with the network of twigs and shadows around her, and she faded into the background.

She scanned the tropical forest understory beyond the maze of branches, but her gaze was drawn back to the limb just before her. Revealed there was what looked like the head of an insect floating in the moss—an oblong visage from which grew paired fronds that could have been antennae above a glossy black eye spot. A body was not apparent behind this head—it appeared to be attached to the finely veined sheath of a fallen leaf, which was overlain with the common blotchy texture of the moldering forest floor. But parallel to this leaf was a long shank ribbed in a herringbone design.

As the margay drew closer to study the scene, she detected the faintest motion—the long appendage had tightened. Her eyes flashed and she pounced, moving faster than she could think, but pinned only moss between her claws.

The hopper had launched reflexively, accelerating to a speed that would cover three meters in a second—faster than a cat's eye could follow. Ten centimeters into its flight, it caromed against a twig. The stunning force of the collision would have incapacitated most creatures, but the grasshopper was built like a bullet with a helmet for a head, and the impact only set it spinning. It tumbled through the air with six legs spread in all directions, so even though it landed awkwardly on its back, it had one foot in position to grasp the bark and right itself. It gathered its grip, cocked its jumping legs, and hurled itself into the air again, all in less time than it took the cat to dive from the branches and pounce on the spot.

The cat stood with one paw held up at shoulder height, her eyes tracking the hopper's crazy trajectory. The insect ricocheted back and forth, disappearing along random paths, somersaulting through the air until it touched down, found its purchase, then sprang again twice, three times. It bounced here and there, finally launching into an arc that would land it in the water. This was an opportunity for the cat, who knew that an insect trapped on the surface is easy prey.

The world spinning in the grasshopper's eyes was dominated by a hemisphere of greatest brightness—sky light. Now a counterbalancing light entered his vision from another direction, with the same hue as the sky but only half as bright—polarized, reflected light. He deduced that he was descending over water. So he opened his wing covers and spread black and orange wings that rippled in the airstream until they slowed his tumbling descent. He rolled level and rose under slower but controlled flight, headed away across the stream.

A jacamar perched twenty feet above the scene heard the heavy insect flying toward the brush on the far shoreline. But before she could take flight in pursuit, the display of bright wings disappeared—as though a branch had intervened in her line of sight. Her gaze tracked the flight path, but the banner of the insect's colors did not emerge back into view. She knew that the hopper must have touched down and folded its wings in the same motion. It had vanished right before her under camouflaged wing covers snapped shut so quickly that she had not been able to mark its location. She considered descending to hunt through the general area, but the proximity of the cat dissuaded her.

At the water's edge the margay stood watching her meal sail out of reach. She turned to pick the driest course back along the shore. But before she could take a step she was startled when the stone she was about to walk on shot off across her path.

The frog had catapulted himself into the air sideways, exposing his position at the last instant before being stepped on. He tumbled along his trajectory, all four legs spread, each one tipped with tacky toes. He flew within a spinning disk of digits, its soft circumference ready to stick to whatever he collided with like mud thrown against a wall.

The cat looked off in the direction of his flight toward a broad leaf that now swayed against the still backdrop of the understory foliage. Eventually she noticed a dark spot fixed to the underside of the green membrane, then recognized the form crouched there. As soon as she focused on it, the frog sprang again.

Where the grasshopper had been rigid, the frog was elastic—right down to his flexible bones. He flattened out on impact, killing his motion. Even though he had landed upside down with his white belly showing, he had one foreleg bent double beneath him—with which he flipped himself over directly into a coiled posture. He still had time to alter his launch direction slightly before bounding out from under the cat's next attack.

The frog landed in an inch of water, but there was nothing for the cat to find in the center of the splash. Already the amphibian had moved off under the surface, swimming away nearly as quickly as he had flown through the air. With every stroke the webs between the swimmer's toes expanded, pushing him ahead. As he brought his knees back up for the next stroke, the webs folded, offering no forward resistance.

The frog swam in a series of pulses, changing direction with each, and when the cat perceived its direction and gave chase, the water she stepped into sent ripples expanding away from her footfall. The ripples made the image of her escaping prey waver—its refracted form stretched, then flattened impossibly as each wave passed; the creature appeared to be moving in two directions at once. She watched in confusion as the distorted swimmer grew indistinct beneath the broken silhouettes of trees reflected on the surface of the deeper water. Pursuit of this hopper, like the pursuit of grasshoppers, would take more energy than it was worth.

The cat was well suited to chase small, short-legged prey—be they small mammals or insects—that ran along the bark. She had inherited genes that endowed her with cunning and quickness, sharp senses, and sharper teeth—traits that had been honed over millions of years of natural selection.

But the hoppers had the jump on her. They had been born with genes for evasive ability refined over a history more than twice as long as that of the feline, and over twice again as many generations—unlike the cat, they reproduce in the same season they are born. Their doubly long legs insured their survival against all the enemies that would appear in their world.

The margay backed away from the streamside and flicked the water off her paws. Then she turned her back and set off to hunt creatures that lived in the dry branches. Over the years, as she grew to understand her environment in ever more detail, she recognized the futility of pursuing the various hoppers. She eventually learned to ignore them all—other than to sometimes feint in their direction and then watch them disappear.

Frogs cannot rise on wings the way insects can, lifting themselves to safety above their predators. Nonetheless, frogs and even some reptiles still depend on escape into thin air as a last defense. In the Old World tropics, tree frogs that live far from the security of water escape through flight. They do not generate lift but gain the height they need—ahead of the time they need it—by climbing trees.

When discovered by a prowling marbled cat (an arboreal feline whose form is broken by curved black and white stripes, the South Asian equivalent of the margay), one frog leaps away into the airspace a hundred feet above ground. As it plummets earthward, the animal spreads its broad, webbed feet and flies with the aid of the webbing its ancestors used for swimming.

When this frog is falling fast enough to feel air resistance between its toes, its feet expand into flat pentagons and its body flattens into a disk. The animal tenses its legs and stands on the air—maximizing the air resistance by holding itself horizontal. Then if it leans to one side, it falls off in that direction, turning its course. Unlike the hopping frog that tumbles along a ballistic trajectory, this creature glides in controlled flight—more controlled the faster it falls.

The flying frog weaves through the branches, sailing at ten miles an hour, selects a vertical trunk that may be fifty feet from its launch point, and lands securely on all four sticky feet. With its ability to control and flatten its flight, the creature saves the energy it spent climbing—energy lost by flightless hoppers that leap and then find themselves free-falling to an uncertain landing.

These same trees are home to the flying lizard. This reptile has extended ribs with a membrane between them. When it leaps from the trees, these ribs open out into wings that span the space between its outstretched front and back legs. The flying surface is attached to the lizard's hind legs, which provide the tension that controls the animal's flight.

Flying frogs stick to the bark when they land, but this lizard lands on the run. It flairs just before touchdown, killing most of its speed by pitching up to land vertically with all four claws snagging the trunk. It does not stop on landing but conserves the last of its momentum by skittering across the branches just as quickly as it flies through the spaces between them. It races into the cover of the denser foliage, then, if pursued closely, it continues along a path that extends out into the air, again one step ahead of whatever pursues it.

Successful strategies draw attention—the flying frogs and lizards of South Asia have grown in numbers and become prey for the flying snakes. The paradise flying snake is a common example—a serpent with emerald green scales along its flanks and berry red dorsal beads. When this creature is confronted with the edge of a precipice high in the canopy, it springs off into thin air. Once airborne, it expands its ribs and flattens out into a ribbon—slightly concave underneath. Holding itself horizontal, it slithers along on the air resistance trapped beneath it. As it falls, its head sways from side to side, and its body follows—it undulates along a sinuous path through space the way a ground snake moves through grass.

The paradise flying snake changes course in midair by tilting its head to one side; its body responds, and it ripples around trunks and vines, finally ending its flight draped over the branches of its landing site. Like the other gliders in the forest, the flying snakes can avoid their feline foes by stepping into the sky. But when these creatures leap for safety, they must remain alert to winged predators that watch from even higher. Those modern dinosaurs still command the top of the forest food chain, continuing their dominance over the other reptiles and amphibians—a saurian legacy that has endured for the past two hundred million years, now in its fully feathered form.

Back in the New World, the margay continues to stalk the understory. She and her clan had driven out much of the prey in the leafy scaffolds high above, and now she walks again along branches that bend down to the streamside. There she sees a basilisk lizard hunting ants along the shore, and she steals closer. She moves to put the lizard between her and the water, to block its retreat into the underbrush. Her intent is to herd her prey into the stream, where the water resistance will slow its flight enough for the larger cat to catch up.

The margay dashes from ambush across the opening, and the lizard reacts as expected, retreating in the direction of the water's edge. It reveals a capacity for sustained speed by standing up on its back legs, not scuttling across the ground on all fours but running like a dinosaur. And as it sprints off into the shallows, it does not slow at all.

The basilisk lizard continues its dash across the waterline and then runs across the surface of the stream just as it ran across dry ground, still striding upright as the water deepens beneath it. Sheaths around its toes expand, broadening its footprints.

The lizard is able to walk on water just as the flying frog can stand on air—the resistance it feels comes from the speed at which its feet fall and compress the medium below. The sprinting reptile slaps the surface with every stride, generating enough back pressure to support itself before its foot sinks. When it pulls its knee back up, the sheaths between its toes close, offering no resistance as it extracts its foot from the water. It sprints at a meter per second across the deep stream center where the cat would have to swim. Finally, the lizard collapses onto all fours, folds its legs back, and swims snakelike through the shallows to the safety of the opposite shore.

The margay spends little time considering the loss of yet another meal but simply returns to her continual hunt, urged on by her continual hunger. She has killed thousands of smaller creatures in her life, but each is a challenge. Clever means of escape have come to be prominent features of all her prey. When the chase grows unproductive, near-starvation drives her to risky migrations through unknown territory in search of better hunting grounds.

While over the ages the cat's prey has grown better at evasion, her own line has grown more lethal in parallel. Predator and prey coevolve in a constant effort to diminish each other's prospects for prosperity. Only the swifter member of each contest is assured of survival, and even then, for no longer than the day at hand.

Science Notes

"Strange country this. It takes all the running you can do, just to keep in the same place." So spoke the Red Queen to Alice (Carroll, 1908). The ecological correlate, the Red Queen theory, states that in an unchanging environment, an organism must constantly evolve just to stay abreast of the competition from the creatures with which it is coevolving (van Valen, 1973). As predators become more efficient, prey find more ways to escape.

Flight, the first line of defense for creatures low on the food chain, is not restricted to animals with wings. Hoppers are creatures with legs that extend long enough to equal their body length. Pushing against their rigid exoskeleton, grasshoppers achieve acceleration from their hops proportional to the distance across which the acceleration takes place (Brown, 1967). Thus, long legs give the biggest kick. The impact of the grasshopper against a twig described in this tale has a force of almost twenty times that of gravity (g); 20 g would crush a creature without an exoskeleton, but the grasshoppers are unaffected. Deployment of their wings only slows them down.

There is little grass in the deep tropical jungle, but the name "grasshopper" has followed the tropical hoppers down into the shadows, where they share the strategy of the hopping defense with a menagerie of other creatures.

The disappearance during landing of brightly colored wings (concealed by cryptic wing covers, or overwings) is an advantage refined not only by the grasshoppers but also by the underwing moths. Insects that hop appear in the fossil record some 285 million years ago in the Permian period; the amphibians came to the fore around the same time, the reptiles in the ensuing era.

The cats separated from the predatory mammalian lineage only about forty million years ago in the Eocene epoch—they are relative newcomers among the older lineages. The margay (*Leopardus wiedii*) is a neotropical species, a leopard the size of a house cat; the setting of its interlude here is Tobago. The marbled cat (*Pardofelis marmorata*) occupies a similar niche in South Asia.

The biggest of the flying frogs (*Rhacophorus*) is Wallace's flying frog (*R. nigropalmatus*), which is big for a tree frog—the size of a leopard frog. It was named by Alfred Russel Wallace, an explorer of the tropics who toured Asia while Charles Darwin was in South America; the two of them made contemporaneous deductions about the evolutionary origins of the creatures they encountered on opposite sides of the Pacific. The gliding creatures of the South Asian forests also include the flying lizards (*Draco*; McGuire and Dudley, 2005) and the flying snakes (*Chrysopelia*; Socha, 2002), such as the venomous paradise flying snake (*C. paradisi*). The basilisk lizards (*Basaliskus*; Hsieh and Lauder, 2004) are creatures of the New World tropics.

References

Brown, R. J. H. 1967. Mechanism of locust jumps. *Nature* 214:939.

Carroll, L. 1908. *Through the looking glass: And what Alice found there.* London: Macmillan.

Hsieh, S. T., and G. V. Lauder. 2004. Running on water: Three-dimensional force generation by basilisk lizards. *Proceedings of the National Academy of Sciences, U.S.A.* 101:16784–88.

McGuire, J., and R. Dudley. 2005. The cost of living large: Comparative gliding performance in flying lizards (*Agamidae: Draco*). *American Naturalist* 166:93–106.

Socha, J. J. 2002. Gliding flight in the paradise tree snake. *Nature* 418:603–4.

Van Valen, L. 1973. A new evolutionary law. *Evolutionary theory* 1:1–30

Nutcracker

Two Hundred Years Ago

In some cases, the symbiosis between just a few organisms holds an entire ecosystem together. Such is the case in the environment at Lost Lake—a bright pool of crystal-cold snowmelt set behind a precipice high in the western mountains. On this day its surface mirrors the ridgeline beyond the canyon to the south—ranks of distant peaks that seem close enough to touch through the thin air. Their frosted crowns are reminders of the cold of the previous night—a chill now fading from the air as the day warms.

The surrounding alpine forest is in full leaf; it is the height of summer. The trees spend the mild, bright days setting their seeds, and the animals around them are busy stashing those nuts away for winter. The rigors of this environment are never far off. This country knows fewer than thirty frost-free days a year. The native animals and plants must endure extremes of fire and ice to persist here.

The steep terrain is too rugged for humans to penetrate, but a year earlier, Meriwether Lewis and William Clark were able to pass through the canyon far below on their journey of exploration to the Northwest. The scene high above their passage—up here by the lake—is framed by a grove of alpine trees. These are stone pines, forty feet tall. Their green scaffolds shelter a fir and spruce understory in which a pair of black-and-white birds perch, their calls echoing across the water.

Brittle thickets of manzanita, chinquapin, and gooseberry extend through the shade at ground level. Their pungent foliage fills the air with the promise that the calm will not last. After decades of steady increase, critical fuel mass has accumulated in the layers of debris below the trees— the potential for fire hangs heavy in the September stillness.

The midday heat on the exposed rim farther up the cliff has become too much for the marmot foraging there. He abandons his search for pine nuts blown up onto the ridge and retreats down through the rocks to wait for the sun to move into the trees. Within a narrow cavern in the talus he hollows out a resting space, digging for the previous night's coolness still stored deep down in the sand.

Airmail. A Clark's nutcracker eyes a whitebark pinecone.
The bird may carry pine nuts for miles before burying them.

As he works, his back feet dislodge a slice of rock that has spalled away from a weathered face. The flat wedge topples onto its side and slides away on loose sand across the floor of the cavern, down a steepening chute, finally falling through the triangular opening between two boulders. The marmot looks toward the space where the rock disappeared. His nose twitches for a few seconds after the noise is gone, then he goes back to his digging.

The thin brick slips over the cliff and falls silently beside the vertical face, its slow rotation violently interrupted when it crashes through the brush at the base of the wall. It shatters against a black obsidian outcrop, sending a shower of sparks into the tangle of deadwood there. A few moments after the silence returns, the settling dust fades to smoke. Flames emerge, climbing a skeleton of weathered branches, the tips of which sag smoldering into the deadfall below.

The ground fire spreads through the driftwood, igniting live summer-dry plants along with piles of bark and cones accumulated since the last fire, twenty-five years ago. The blaze is not driven to rapid advance until the afternoon breeze guides it upslope into a slanted clearing overgrown with more bleached tinder. Mice and ground squirrels retreat toward their burrows at the scent of smoke, but as the crackle of flaring brush grows louder in their ears they spin about and run across the wind as far as they can, abandoning their homes.

The wall of fire rushes above the low growth on a steep rockslide, its momentum building ever-higher temperatures, its course random—shifting with the caprice of the breeze. The strength of the conflagration doubles, and doubles again; the flames follow the contours of the fire-spawned wind around the curve of the ridge, then back toward the lake.

When it reaches the trees, the fire is hot enough to keep burning right on through them. Its leading edge dries the last of the moisture from already desiccated needles, and its energy builds as each crown bursts into flame, one preheating the next.

Orange sheets of fire explode through the branches from tree to tree. Snaps of single needles popping into flame are multiplied by the millions into a roaring crescendo that kills every thin-barked fir and spruce it touches.

The firestorm climbs all the way up to tree line, where its incandescent front rises above the crest of the ridge. Wraiths of floating flame fold up and consume themselves as they rip away from the burning rim and fly out over the far side; below them, flaming ingots roll down the rocky wall. But no more fuel waits on the steep, glacier-polished opposite face of the ridge, so the fire stalls and burns lower. It casts its embers to glide out over the valley beyond—the longest lived of them ready to start new spot fires where the wind finally sets them down.

The wildfire leaves a path of desolation for thousands of feet up the mountainside. Spring rains turn the dark scar vermilion with a flush of fireweed—its dormant seeds germinated by the coating of ash. Other seeds fall into the area, including a few fir and spruce nuts borne on their single wings, delivered by winds flowing upslope from the lower-elevation forest. But the seedlings they produce do not survive the drought of the following summer—they collapse in wilt on the burned, exposed slopes and are dead before the fall's first snow.

Nonetheless, fire scars accommodate the growth of one guild of alpine trees that is adapted to life in the ashes. These are the fire pines. Their seeds prosper at the stage of the fire cycle that immediately follows the devastation. Fire-pine seedlings are slow growing, spending more of their energy putting down roots than putting up tops. Their diminutive crowns demand less water, allowing them to survive the drought of their first summer on unsheltered exposures. They thrive in the absence of competition on unshaded faces or charred soils. They are pioneers, trees that colonize barren terrain, extending the forest margin, pushing up tree line.

The lodgepole pines from the shallow meadow margin, the bristlecones on the ridge, are fire pines. The fire kills these trees, but their seeds survive. Unlike the nuts of other pines, these seeds are not shed into the soil each fall. Had they been cast away each season, most would have fallen prey to the rodents and birds who constantly scratch at the carpet of fallen pine needles. The few soil-borne nuts that chanced to pass undiscovered on the ground for the decades between fires would have been killed by the passing inferno.

But the fire pines store their unshed nuts above ground in sealed cones fixed to their branches. Much of the heat of the fire that roasts those cones is absorbed by the melting of their coating of pitch, protecting the dormant seeds within from the highest temperatures. The molten pitch drips the heat away as the flames pass and, with their lacquer seal broken, over the following days the cones open after years of waiting.

So after the firestorm, the marmots found it snowing pine nuts. Seeds twirled in the air, their wings lifting them off flat rocks with the slightest breeze. Their descent was arrested long enough by their autorotation to maximize lateral flight before they came to rest. Unfortunately, the winds upon which these seeds depended for their dispersal blew steadily away from the fire scar during the weeks of fair weather that followed, and the fire-pine seeds were carried off in the opposite direction. They floated past the lake and up over the ridge, leaving the burned area unseeded.

Nonetheless, this pinewoods would regenerate. There was a pine here that would take advantage of a space where nothing else could establish itself. This pine alone would be able to profit from the cache of liberated nutrients now waiting in the ash. Its parent trees had stood far enough away to avoid the fire. Yet their boughs were ready to launch their wingless nuts from a long distance, and from many directions. Those nuts would fly to the burn site against the wind. The seeds were large enough to sprout hearty seedlings, able to send down roots into the rocky soil faster than the late-season water could run away. They would eventually provide the shade that would permit the return of the mixed woods to the site and continue the forest's cycle of succession of tree species. These vanguard trees were stone pines.

The stone pines are found across the north temperate regions of the world. In Eurasia they grade from Swiss stone pine in the Alps through Siberian stone pine in the Urals to Korean stone pine and across the East Sea to Japanese stone pine. They are found in rugged, high-elevation terrains on stony ridges at tree line. Their seeds anchor the diets of many pine-forest creatures: red crossbills, pine grosbeaks, finches, and jays; chipmunks, squirrels, mice, moose, and elk. Stone-pine seeds make up a large part of bears' diets, though bears don't climb trees to get them. The bears wait for pine squirrels to cut down cones and collect the seeds in middens. As fall approaches, the bears raid these caches—to the disconsolation of their smaller owners.

The sole North American stone pine is the whitebark pine. It establishes best in recently burned areas. Its cones are not closed and dependent on heat for their opening, as are those of other fire pines. Stone-pine cones, almost as fat as they are long, have scales that mature partially open, though neither wind nor fire will release the seeds. Those seeds stay put until picked out one by one. The scales bristling from these cones have a particular weak point—the end of each can be broken off, exposing the pair of seeds sheltered beneath.

Those scales are broken, and the seeds removed, by nutcrackers. Each stone pine supports its own subspecies of nutcracker. The nutcrackers of the mountain provinces from the Carpathians to Kamchatka are dark, spotted birds of the crow family, with long, sharp bills perfectly shaped to shatter pinecones. They thrive in the high mountains, well above the zones dominated by cedar and the harder-wood pines.

The single species of nutcracker in North America is Clark's nutcracker. This is a monochrome bird—black and white and gray. Its bright wing patches and outer tail margins flash in summer against the deep green of the pine woods; its black primary flight feathers and bill contrast with the white winter later in the year, when most other creatures have abandoned the snowscape. Its *kra-a-a-ck* call is muted by the softness of the snowy woods, but in summer it carries—bouncing off the rocks.

That call filtered downslope through rugged talus-filled canyons in territory that would one day be named the Bitterroot Mountains and attracted the attention of the passing members of the Lewis and Clark expedition in 1805. The name of one of the trip leaders was eventually bestowed on the nutcracker in honor of his descriptions of it. (The other trip leader, Meriwether Lewis, gave his name to a woodpecker.)

Clark's nutcracker, which feeds solely on pine nuts, has a special relationship with the whitebark pine. The cones of the whitebark pine do not hang pendulous from the branches, as do cones that dispense seeds onto the wind—like the Jeffrey and sugar pines. Whitebark cones protrude flat out from branch tips, making them easy landing perches for nutcrackers. Clark's nutcracker readily fractures the scales on those cones and retrieves the large seeds from within.

These seeds are not winged—they are dispersed independent of the wind. The nutcracker itself transports them to germination sites. The wings on lodgepole- or ponderosa-pine seeds can extend the distance those seeds fall to hundreds of yards; Clark's nutcracker can carry whitebark-pine seeds ten miles. With this aid to their transport, those seeds can afford to be heavier—among the heaviest of pine nuts—the better to support a first-year seedling on the southern aspect of a burned-over moraine.

Heavy, rich stone-pine nuts alone provide all the sustenance a nutcracker needs. The birds depend on them year round, though the nuts are available in the cones only during late summer. The nutcrackers extend their nut-feeding season by storing the excess of the crop. They carry pine nuts and search for cache sites, where they bury them for later, winter retrieval.

Since it lives on stored pine nuts, Clark's nutcracker can raise its young any time of year. The birds choose to nest in winter, when competing birds have retreated to the lowlands and predators of nests and nestlings have gone south or into hibernation. Nutcracker chicks are raised in the gales of February on a pure pine-nut diet.

The stone pines and the nutcrackers thrive in this habitat through their interdependence. The trees grow only where the birds plant them—seeds left in their cones die in place. Seeds cached underground delay their germination, not sprouting with the spring thaw but waiting until the heart of the summer. The nutcrackers can rely on their hidden stores to remain firm from October through July, by which time the next season's crop will come ready.

The nutcrackers hide late summer's pine nuts far upslope, working until the higher elevations are buried in snow; then the birds move downslope. Once the cones are emptied in the fall, nutcrackers begin to retrieve their lower-elevation reserves, waiting to recover the rest as the snowline retreats upslope with the spring thaw. These birds deftly recall all their cache sites, uncovering them as each is revealed in turn with the evolution of the new season. Late in the spring, the birds may dig through the snow in their impatience to get to sites not yet exposed.

Some birds experience nesting failure and don't retrieve all their caches; others are taken by goshawks or merlins and don't retrieve any. Those buried seeds will eventually germinate. Cached above tree line, they will grow stunted—those at the highest elevations will struggle to provide even a low ground cover. Seeds stored away late in the season at lower elevations beneath aspen or ponderosa pine will germinate in shade and die—unless a fire should burn away the forest above them.

But seeds planted in burn sites prosper, and nutcrackers have an affinity for burn sites as cache locations. They will carry whitebark-pine nuts for miles to reach such sites. Seedlings that germinate on ashen, shade-free slopes will thrive under the full ultraviolet intensity of the high mountain sun.

After the ashes had cooled, Clark's nutcrackers appeared at the burn site on the shoreline of Lost Lake. The birds hopped among the boulders, probing the soil, stashing whitebark-pine seeds in lots of four or five an inch below the surface. After each effort, the birds stood back and inspected their work, glancing around for thieving jays or marmots who might conspire to steal their cache. Then each flew off to refill the seed pouch below its bill— the equivalent of the chipmunk's cheeks—and returned again. Each nutcracker worked all day as the summer lengthened, caching fifty thousand seeds or more in a season, in thousands of different places.

The mix of trees in the temperate forest is in a constant state of flux, reacting to the instability of the landscape. Steep walls lose their footings and slide into valleys, damming the drainage and creating lakes. These lakes are temporary—they fill up with stream-borne sediment until the bottom reaches the surface and they become marshes. The marshy shore first sprouts willows. Farther behind the shore come aspen where the swampy ground consolidates, then lodgepole pine, and finally ponderosa or Jeffrey pine. Each species advances on the next, displacing its predecessors as the ground grows firmer and better drained.

On the drier ridges above, fire scars or landslides will first be colonized by shade-intolerant whitebark-pine seedlings. These pines are eventually replaced by an advancing wave of fir and spruce, whose seedlings survive in the shady, sheltered understory below the pines. The pioneering stone pines grow tallest on the sunny edge of the advancing forest, pursued by a wave of faster-growing, more shade-tolerant species that will eventually overtake them—until the fire surges through again to overtake the lot and reset the cycle.

The whitebark pine is a relatively young species in North America. Its ancestors appear to have come from a stone pine–rich Eurasian source. They were carried as seed by the ancestors of Clark's nutcracker across the seldom-seen province of Beringia—the land-bridge connection between Siberia and Alaska that appears during periods of low sea level.

The crossing will have taken many pine generations, during the height of an ice age past. Seeds cached along the route would have germinated and grown, leaning into the polar blasts that swept down from the frozen Arctic Ocean. Saplings there would have adopted the low, ground-cover scrub habit they now employ high in the Rockies at tree line—in a modern environment similar to that of the long-gone low-lying land bridge.

The low-growing stone pines of Beringia were inundated by the reopening of the Bering Strait during the interglacial warming. But the easternmost of them had reached North America by then. They moved into the mountains as the glaciers moved out, and spread south—carried by nutcrackers through the Canadian Rockies and into the mountains of the future United States.

Since then, the whitebark pine and the rest of the American white-pine family (five-needle pines—the bristlecones, the foxtail pines, the sugar pines) have grown distinctly different from their Asian forebears. The isolation of the Eurasian white pines from those of North America has led to divergence between the two communities, and between their respective allied cohorts of birds and mammals, understory plants, and microbes. One seemingly minor member of this transcontinental divergence is a fungus—Eurasian white-pine blister rust. The rust is endemic in Eurasia, and the trees there are acclimated to it and resistant to it.

One Hundred Years Ago

In an operation that was perfectly innocent in the context of the times, the forestry industry of the late 1800s made a series of missteps that precipitated a disaster. American white pine—a lumber species—was exported to Europe in a bid to enhance the timber industry there. The transplanted North American trees had not encountered white-pine blister rust since their ancestors left Asia in the prehistoric past. They were no longer acclimated to the pathogen, and were so devastated by the endemic fungus in Europe that the transplantation experiment was abandoned. But before it was, American white-pine seedlings were exported back to the United States on the mistaken assumption that seedlings infected abroad could be detected and culled.

The centennial of the Lewis and Clark expedition brought bad tidings for the diverse forests the explorers had described in close detail. That year saw the first American case of chestnut blight, an introduced fungal disease that would spread implacably to destroy the American chestnut in the lowland woods of the continent. Also discovered in 1906 was the first American case of white-pine blister rust.

Blister rust follows a complex life cycle whose early stages mature on alternate hosts—currants and gooseberries. In areas where those understory plants were prevalent, the introduced rust spores spread through American forests on the wind before anyone thought to martial the quarantine necessary to stop them. The fungus dealt a devastating blow to the indigenous white-pine timber industry. In the decades to follow, half a billion wild gooseberries and currants were pulled up in an effort to thwart the introduced pathogen—an effort that came to naught.

The magnificent vistas across the slopes of the Sawtooth Mountains, the Wasatch Range, the Bighorns, and the Canadian Rockies were once framed by stands of whitebark pine rising almost fifty feet tall, eight thousand feet up in the thin air on the approach to tree line. Most of those trees would die or become irreversibly infected with white-pine blister rust. Whitebark pine is one of the most susceptible of the American white pines to the introduced fungus. In the area through which the Lewis and Clark expedition passed, late summer conditions of coolness and humidity and a vibrant growth of gooseberry in the understory conspired to create perfect conditions for the spread of the rust. The disease will lead to the extinction of the whitebark pine there.

A potential salvation from the fungal disease appears in some infected stands of whitebark pine. As the blight burns through an area, a single tree occasionally will be left untouched, standing strong and green among hundreds of its moribund peers. These sole survivors are endowed with a genetic resistance to the infection; they embody the best chance of recovery for their species. Their seeds have the potential to replace the susceptible trees with a new generation of saplings that will withstand the fungal onslaught.

The Eurasian stone pines that coevolved with the fungus have established a stable coexistence with blister rust. The precipitous appearance of Western culture in the New World has led to extensive fungal infection of the American trees in a relatively much shorter period—a time that has provided other complications as well for the whitebark pine.

The settlers who moved west following the Lewis and Clark expedition were aghast at the free-running forest fires that regularly regenerate the pine forests. They did not comprehend the place of fire in the natural order—all they saw was destruction, so they fought the fires that arose naturally in the fall, arresting their progress before they could expand.

That practice curtailed the prevalence of burn sites in which whitebark-pine seedlings could become established. Light gaps in the forest canopy were closed over with faster-growing fir and spruce species no longer regularly thinned by fire. Whitebark-pine seedlings grew scarce in the managed lands.

Dense fir and spruce cover fosters epidemics of mountain pine bark-boring beetles, which attack and weaken mature whitebark-pine trees, curtailing nut production. Reimposition of the fire cycle would allow the seeds from the few remaining resistant trees to germinate and replenish the whitebark-pine populations, but civilization had encroached so far into the forest that even controlled burns posed an unadvisable risk for structures.

So the rare seeds that carry the hope of a new generation of trees resistant to the fungal blight are prevented from growing. The nutcrackers deprived of their stands of whitebark pines cannot find many nuts—they consume their entire stash, barely able to feed a few nestlings—leaving very few seeds to germinate. Those seedlings that do break the soil find themselves unable to reach the light, buried in the shade of the faster-growing firs and spruces. At the same time, the surviving parent trees suffer intensified attack by bark-boring beetles.

The Present

On the second centennial of the Lewis and Clark expedition, Lost Lake is surrounded by an impenetrably dense growth of fir and spruce trees. Only their own shade-tolerant seedlings can survive in the deep shadows they cast. The skeleton of a lone whitebark-pine snag sags out toward the edge of the ridge beyond the lake, its brown needles still attached.

The call of the nutcracker echoes across the water only in memory. The entire interdependent community sustained by whitebark-pine nuts has disappeared—destroyed by a human act, even though no human has ever set foot on the inaccessible shore of Lost Lake. It will take a significant effort to restore what was so easily undone.

September dries a reservoir of dead brush and papery leaves that accumulates deeper and deeper behind the lakeshore. The wildfire that should have cleaned it away and thinned the firs and spruces is long overdue. Chances for release of that fire remain, of course, and grow with every passing year. The initiating flame will eventually kindle spontaneously, its potential growing too great to suppress. The inevitable forest fire will expand into an uncontrollable monster—larger the longer it waits to be released—crashing across the landscape like a tidal wave, flying across rivers and canyons, burning for weeks, laying waste to thousands of acres, consuming all the ground cover, and then the soil itself.

And in its aftermath, there will be no whitebark-pine seeds to reestablish the forest at its burned edges, so the wooded margin will retreat. Formerly root-bound soils will become unstable in the spring runoff; tree line will come down, as will the pyramid of species that once thrived in that transitional habitat, foraging for whitebark-pine seeds. The bears, which once spent the fall feasting on pine nuts, will be driven downslope in search of the forage to fatten them for hibernation. Their lower-level prospecting will push them into conflict with humans, who have been building their own dwelling places upslope into bear territory.

Below the ice-bright morning sun, the view east from the peaks of the southern Sierra Nevadas spreads away over a folded landscape. A receding succession of low, north-south mountain walls divides Nevada into dusty, bone-dry valleys—the basin and range province. Water flowing down from the Sierras, or from the Rockies far to the east, does not find its way to the sea from this area—it never emerges from this Great Basin but disappears into dry washes and alkali flats. Distant blue mirages provide the only hint of water in this desert, shadowed by the vultures wheeling overhead.

But the tallest of the low ridge crests pierce the cloud ceiling and wring an occasional mountain shower from a sky that otherwise rarely rains. These tallest crowns harbor sky islands—clean, crisp oases of green floating above the miles of baked desolation far below. The largest of these shady refugia support full forest cover, and there whitebark pine survives.

The sky island climates in the Great Basin are too dry to support rapid fungal growth, so the spread of white-pine blister rust is curtailed there for the time being. These island outposts are like the lost, Old World mountain refuges where populations of such ancient lineages as the dawn redwood and the ginkgo survived to modern times. Those living fossils were isolated in their mountain redoubts from conditions that had long since driven the majority of their kind to extinction. The sky islands of Nevada provide similar refuge for the whitebark pine, while its primary populations to the north fail. The call of Clark's nutcracker still carries through these isolated trees. Their boughs cradle a reservoir of whitebark-pine germplasm that remains available for use in a mountain pine reforestation program.

This recovery program will have to be telescoped in time. It took eons for white pines to coadapt with blister rust in Eurasia. The American whitebark pines do not have that long. Fungus-resistant trees will have to be propagated and then replanted, even though the fungus continues to mutate—new strains may arise to infect trees that were resistant to previous forms of the disease. Many people will have to mount a sustained effort to succeed in repairing the mistakes of the past.

That effort—to restore the whitebark-pine forest and preserve the associated wilderness—is already under way. The first steps will be small—the cultivation of rust-resistant tree lines, the conservation of whitebark-pine habitat, the reinitiation of the fire cycle. But from that point, the reoccupation of the subalpine slopes by whitebark pine will be self-sustaining—the trees are already well adapted to fill that niche. They had been there for millions of years before the coming of humans. And their nutcrackers are still close by, ready to assist in the effort.

Like the growing potential for a forest fire in a long-unburned woodlot, the fungus-resistant trees will rise again on the fungus-cleared ridges they are best suited to colonize. The pioneering seeds will be brought by Clark's nutcracker, returning to its favored food source after decades of imposed exile among alternate hosts, such as piñon and Jeffrey pines. Over a span measured in the generation times of pine trees, tree line will be raised again to expand the forest back across its upper margins. This restoration of the subalpine forests of the Rockies, the Cascades, the Bitterroots is a goal hundreds of years away. Nonetheless, it is being pursued today by men and women who have a clear view of our past mistakes, and a constructive, optimistic vision of our future ability to restore the lands of the West.

Science Notes

Whitebark pine (*Pinus albicaulis*) is a member of the five-needle pines, along with white pine, sugar pine, and bristlecone pine. Jeffery pine is a three-needle pine; lodgepole pine is a two-needle; piñon pine is a one-needle (needles not bundled). Whitebark-pine seedlings do better in recently burned sites, but they do not use the serotinous (heat-activated) cone strategy other fire pines use to make seeds available when such locations become open. Instead, the whitebark pine has a symbiotic relationship with Clark's nutcracker (Tomback, 1998), which plants the tree's seeds (Lanner, 1996; Tomback et al., 2001). The nutcrackers are in the same family as the jays and crows—birds known to stash seeds away during times of plenty. The natural range of whitebark pine includes the northern Rocky Mountains, the Cascades, and the ranges between them. It grows at higher elevations down through the Sierra Nevadas and across the higher ridges of the Great Basin.

This pine has been debilitated by blister rust and has had its habitat altered by suppression of the fire cycle (Tomback et al., 2001). Fire suppression favors the growth of faster-growing (fire-intolerant) firs and spruces, which shade out seedling whitebark pines. Fire suppression has decreased the frequency of fires but increased their severity when they finally do kindle (Donovan and Brown, 2007). Full stands of fir and spruce (which, as opposed to the climax forest species, do not shed their shaded lower branches) fill the air space beside the bare trunks of the tallest climax forest trees. When these understory trees are dried at the end of summer, fires can ladder up into the tops of the tallest trees of the forest, starting a crown fire that can grow hot enough to melt, then ignite, composite asphalt shingles. In the absence of the dense fir/spruce understory, the fire may be restricted to the forest floor.

The restoration of the whitebark pine and its habitat will be a multigenerational effort. People working to that end would welcome your support. Join them at the Nature Conservancy (*nature.org*) and the Whitebark Pine Ecosystem Foundation (*www.whitebarkfound.org*).

References

Donovan, G. H., and T. C. Brown. 2007. Be careful what you wish for: The legacy of Smokey the Bear. *Frontiers in Ecology and the Environment* 5:73–79.

Lanner, R. 1996. *Made for each other: A symbiosis of birds and pines.* New York: Oxford University Press.

Tomback, D. F. 1998. Clark's Nutcracker (*Nucifraga columbiana*). In A. Poole and F. Gill, eds., *The birds of North America*, no. 331. Philadelphia: Academy of Natural Sciences; Washington, D.C.: American Ornithological Union.

Tomback, D. F., et al. 2001. *Whitebark pine communities, ecology, and restoration.* Washington, D.C.: Island Press.

Flying Lessons

The peregrine falcon was on his yearling migration—the first time he had set out alone to span the heart of North America. He was headed for Canada, the call of the mating grounds drawing him north into weather older birds knew enough to wait out.

There were no thermal updrafts rising from sun-warmed flats today. Instead, squalls of rain divided the sky, hiding the plains horizon behind their dark skirts. But a steady southerly breeze coursed under the low cloud ceiling at twenty miles an hour—a river of humid air flowing up from the Gulf of Mexico, adding its speed to his own. Even though the ride was bumpy, he could handle it—and the tailwind doubled his cruising velocity.

He still wore streaks of brown juvenile plumage, still had much to learn about his world, but he knew about the wind. He had learned to flow with the jarring buffets of his hundred-mile-an-hour power dive, guided by nearly closed wings to impact squarely against moving targets. The swirling wind was his ally, the avenue he always followed.

Low grasslands scrolled past below at forty miles an hour, but when he looked ahead, the falcon found his path blocked. Squarely before him a curtain descended from the base of the cloud deck—a rogue downdraft of wind and rain. Yet the falcon knew that these currents all worked together, so he stayed with his tailwind to see how it would bend.

The descending wall cloud loomed before him, but his river of air responded by shifting upward, carrying him higher. This was a welcome turn of events—he lived to gain altitude. When he achieved the heights, he could float to his destinations in an effortless, shallow glide. He usually climbed the sky on the lazy carousel of a fair-weather thermal updraft. But now his path was straight north and slanting ever more upward—a shortcut to the culmination of his ascent.

As he rose, he could feel the power of these winds growing. He was expending less effort now—nearly coasting—yet he was ascending at four thousand feet a minute. He would be at the cloud ceiling above him in moments. He was looking up into the bottom of the great cumulus cell that drove the winds around him.

The entire understory of the cloud was rotating; rising humidity was

Quite a ride. A peregrine falcon fights against the tempest in a thunderhead.

condensing into rain far above, and above that into snow. The change from invisible water vapor to liquid or to solid ice crystals liberated heat. The amount of precipitation spawned by this one cloud was massive—generating enough heat to raise pillars of white all the way to the stratosphere. The growing cumulus tower pulled ever more ground-level vapor up in a continuous column, and now it was pulling the falcon up as well.

He was rising at more than fifty miles an hour through the nearly vertical updraft, while the downdraft that drove the wall cloud descended at the same speed close in front of him. A few yards away he saw rain smeared into streaks that appeared to be falling at twice his speed. The two opposing drafts were twisting each other into barber-pole spirals, adding a sideways slant to his ascent.

The close-by wall in the sky where the downdraft passed his updraft roared continually where the opposing sheets of wind tore at each other along the virtual partition between them. Their revolution around one another was linked to the rotation of the ponderous gyre he could see in the base of the cloud above.

And then, as the opposing up- and downdrafts continued to strengthen, their building pressures formed the air between them into a new shape, which smoothed the boundary where they met. In the blink of an eye the cacophonous ripping of the winds ceased in the falcon's ears, replaced by a dead-steady whistling tone. Horizontal cylinders of wind had snapped into place perpendicular to the wind-shear direction like the rollers on a conveyor belt, minimizing the friction where the torrents of air rode past each other. No longer dissipating their energies in turbulence, the opposing up- and downdrafts accelerated to yet greater speeds, spinning the transparent rollers between them ever faster.

The song of the wind-whistling slipstream evolved, stepping down in frequency. One by one the rolling horizontal cylinders of air fused, further minimizing the conflict where they spun against each other, creating fewer but larger spinning rollers. For the first time, the falcon wondered about his ability to handle anything the wind could blow his way. He was ascending effortlessly—not flying at all, drawn along as a passenger on the ever-strengthening updraft. He held his wings outspread just to stabilize his body. He had been borne high enough already that the plains spread out far below him, and he was ready to drop out of the updraft and resume his northerly flight.

Finally, the descending tone of the whistling winds before him settled into a preternaturally deep, continuous moan. The spinning cylinders had coalesced into a single massive column that drew its energy from the crease where the opposing updraft and downdraft pressed most closely against each other.

And there, as the falcon looked ahead, this spinning cylinder of air emerged into plain view. A short tube of cloud wider than his wingspan appeared above him. Its rim rotated more rapidly than the eye could track. Its axis was aligned toward him—he could see through its hollow center. In seconds it had lengthened and widened—new wisps of rotating cloud materialized to advance the leading edge of the tube earthward.

The spinning tunnel cloud darkened as it propagated toward him, its width holding constant, but the light in its center fading to black. Just before it reached him, it bent, revealing that the updraft he rode was beside the powerful whirlwind, though very close to it. His ascent was now carrying him skyward at speeds he had previously experienced only while diving. The raw strength of the dark blur spinning before him created a vacuum that was sucking him toward its vortex.

The diagonal funnel cloud curved to extend toward the earth, bending directly below him, depriving him of the option of bailing out by turning around to plunge against the wind. As he strained against the force that drew him into the spinning twister, his attempt to break away was balanced by the inward pull, leaving his course straight up.

The falcon punched through the base of the cloud deck at seventy miles an hour into a weightless, formless whiteout. His sense of direction vanished. He lost sight of the funnel cloud, so he turned and dove blindly away sideways into the mist, only to fall deeper into the roiling heart of the massive thunderhead.

His course soon took him into winds that arose from every side, threatening to flip him over, the sound of the gusts building around him as he flew. He fought to control his flight, remaining upright until he crashed into a wind-shear boundary and was smacked so hard by the opposing gusts that his wings slapped together above his head. He was rolled over and over, pummeled by blasts so strong he was forced to surrender to the elements and fold his wings to protect them from breaking.

He somersaulted through the rushing clouds, accelerated back and forth by collisions with furious walls of air. Finally the buffeting relented and he felt himself coasting free from the assault. He was floating, still in a powerful updraft—sailing along blindly through the mists at eighty miles an hour but with no notion of speed or distance, or of the direction he was being carried. He felt no flying speed—only momentary pulls and pushes from the wind that buoyed him along.

Rain materialized around him—round drops suspended before his eyes, floating along in formation with him but now growing so large that they jiggled in the gusts. He was drenched, continually swallowing the water flowing in through his nostrils.

A shiver shook him, and he realized how cold the air had grown. The raindrops turned white; he felt them clicking against his bill. He shook himself, and the dew on his feathers sprayed away in an explosion of rime ice that was flung back into his eyes by a wave of even colder air.

The falcon's wings flinched open with every invisible blast that burst against him, but he soon learned to hold them closed to keep himself warm against the wintry gusts. His ears popped. The fogs swirling around him thinned, and he flew toward the brighter direction, but his flying produced no sense of acceleration—he was embedded in winds stronger than he could fight, free-falling in directions he could not detect.

Closing his wings no longer had the effect of plunging him into a dive, but he closed them nevertheless to relieve their fatigue. He drifted through space, losing track of time. Focused inward, shivering against the cold, he fought down the nausea that grew from the random spinning.

Finally a brightening in the air refocused his attention to his surroundings. The clouds took form and substance again as he broke out of the grayness into an immense cavern, its far walls thousands of feet away. His wings reflexively spread open into their gliding stance in response to the illusion of great height. The billowing cloudscapes flashed from dark on light to light on dark as lightning arced behind them.

He floated weightless, eyes wide, and stared across the vast space. Shards of mist drifted in its center. They stretched wider, growing relentlessly toward him. He turned to fly from the advancing wall, but his efforts had no effect, as though he were flying in place, making no headway. He pulled in his wings to dive away, but the effort was no use. He was powerless against the unseen drafts that carried him. He felt the air grow still more chill as the vista disappeared into the blinding advance of the wave of snow.

Through the mists he saw the clouds darken above him. The sky expanded into a distant, textured plain as the nearer shards of gray parted. He stared up, fascinated, as the textures grew broader and more familiar until he realized he was looking at the ground. He flipped over and dove for it, accelerating into a plunge that brought the landscape rushing toward him.

The falcon shot through the base of the cloud deck while the curtains of snow around him melted into sleet. The ground was hurtling up to meet him, faster, he realized, than he could control—he was being driven downward at ten thousand feet a minute. He must be in a downdraft, which in a few seconds would drive him into the earth like a falling spear. When he flared his wings to flatten his trajectory he felt the strength of the burst of wind that was hurling him down. Sinking flat through the sky he fought against the downward pressure, trying to stay level, the energy of the elements again whistling in his ears.

The pressure from above diminished as he crossed the boundary of the downdraft and entered quiet, stable air. Off to the side he saw the tornado a mile away descending from the cloud base. It was immense and black, with debris levitating from the earth at its base. Though the cylinder tracked only slowly across the ground, the air within spun faster than a peregrine can dive. Obstacles in its path were smashed by winds spinning at three hundred miles an hour in a pillar of destruction that approached across the prairie at less than ten miles an hour.

The centrifugal force hurling those winds outward was counterbalanced by a vacuum-like low pressure within—an air pressure far lower than at any other point on the surface of the Earth. The opposing forces in that whirlwind, driven by the energies of the surrounding up- and downdrafts, would hold the tornado together against its own self-destructive fury for over an hour.

The falcon did not recognize the terrain he now descended toward—there was no sun direction; he had lost north. He was hungry but unable to see any prey aloft in this weather; sore but uninjured; fatigued but unable to see trees to perch in.

He finally landed in a clump of bushes too close to the ground for safety from the wolves, but at least low enough to avoid much of the wind. He clamped his talons around his branch and closed his eyes against the rain. He would not fly again until the sky was blue, or until he saw other falcons aloft—older falcons who knew what he was learning today about when to fly, and when to wait for flying weather.

Science Notes

The life of *Falco peregrinus*—its movements, physiological demands, and interactions with its own and other bird species—is poorly understood. Juvenile mortality in peregrines approaches 80 percent (White et al., 2002). The falcon in this story is a member of the subspecies *F. peregrinus tundrius*, which pursues an annual peregrination from winter in Central and South America to nesting grounds in Canada; other subspecies are less migratory. This story is set in the past, as the Midwestern peregrine-falcon population is extinct. Efforts are under way to restore that population (e.g., at *www.peregrinefund.org*).

This tale offers a bird's-eye view of the updraft/rotation model of tornadogenesis (Church et al., 1993). The nimbus-diving scene is informed by the account of Lt. Col. William Rankin (1960), who bailed out of a disabled jet at forty-seven thousand feet into the anvil top of a cumulus cloud and spent the next hour being thrown around the sky, eventually coming down sixty miles from his drop-off point.

Mesocyclones, which spawn tornadoes, are found on plains where humid marine air masses collide with cold, dry polar air in South Africa, India, and North America. The spinning funnel cloud becomes visible because of condensate that forms within it from water vapor in clear air that becomes chilled. It is chilled by the low pressure inside the vortex. The pressure falls, according to Bernoulli's Principle, because of the high (cyclic) wind velocity; the relation between pressure and wind velocity can be written as $P + 1/2\ V^2 = K$. This shows that as wind velocity V increases, pressure P must rapidly decrease (K stands for a constant value). Wind temperature falls with decreasing pressure according to the isentropic relation, written as $T = K P^n$, which shows that temperature varies directly with pressure ($n = 0.29$ for air). Thus, as the pressure falls, so does the temperature, leading to condensation, which in turn makes the tornado visible.

References

Church, D., et al., eds. 1993. *The tornado: Its structure, dynamics, prediction, and hazards*. Geophysical Monograph Series #79. Washington, D.C.: American Geophysical Union Press.

Rankin, W. H. 1960. *The man who rode the thunder*. Englewood Cliffs, N.J.: Prentice Hall.

White, C. M., et al. 2002. Peregrine Falcon (*Falco peregrinus*). In A. Poole and F. Gill, eds., *The birds of North America*, no. 660. Philadelphia: Academy of Natural Sciences; Washington, D.C.: American Ornithologists' Union.

Part 3. Sea and Shore

Highway to the moon. Sanderlings depart the Arctic when the evening air freezes. Airborne ice crystals spread moonlight into haloes across the sky.

Sanderling

Over the course of a year, the sanderling was constantly on the move—in pursuit of an endless summer. She flew north along the coast ranges of the Pacific Northwest in the spring, then south along the Atlantic seaboard in the fall. She cut across the Isthmus of Panama in October and headed for an austral summer to be spent on the beaches of the west coast of South America.

She was guided by the stars at night, and by the polarization of the sunlight in the sky by day. She knew landmarks—she had memorized them as a yearling, following in a flock of her seniors. She knew the weather, the humidity, and the seasonal angles of the sun. She was accustomed to the temperatures that presaged the changes that spurred her farthest flights: if it was colder than forty degrees at dawn in a Canadian fall, or warmer than that in the Mexican spring, then it was time for her to move on.

When there was nothing to see—in the dark, or in the clouds over water—she followed the Earth's ethereal lines of magnetic force, landmarks as permanent and reliable as mountains and shorelines. She also knew the scents carried by the wind—the north woods on May Day, the Sargasso Sea at the autumnal equinox—olfactory confirmations of her navigation, borne on the breeze.

The midday sun culminated in Arctic summer, pausing at its high point on the solstice. Her annual cycle reached a high point then as well—she paused along with the sun and made her nest at her farthest northern point before resuming her trek southward.

She sprinted up and down along the margins of waves, then stood in the sand, with willets and godwits foraging in the water downslope, and killdeers and gulls looking down on her from the high-tide strandline farther up the beach. But she ignored the other birds and watched only her own species, waiting to see if her flock was going to take off, and in what direction. In everything she did, she mirrored the other members of her group. They offered each other protection. Foraging or flying alone would be risky for birds as small as they are—the smallest of the sandpipers.

None of the members of her flock knew where they were going when the impulse for flight struck and they followed each other into the sky. The flock had a group mind that all of them followed but none of them led. When they burst from the sand, calling in unison above a waterfall rush of twittering wings, they all watched to see whether the group consciousness would bend their course along the migration route.

She was always ready to go that way, should the flock shift slightly in that direction. But she was slow to go the opposite way, should they find themselves heading out to sea or upriver. With these subtle nudges or hesitations, by her and by each like-minded bird around her, the flock unerringly found its way. They averaged out random daily sidetracks, and over the course of the year, they remained on course to circumnavigate North America.

Earlier in the spring, the sanderlings had been on northern migration up a broad inland arm of the sea along the Canadian coast. Deeply forested mountains rose out of the water on both sides of the strait. The steep slopes disappeared into mists above the mainland shore to the east and over parallel shores of long islands to the west.

The flock was flattened close to the water, where wind resistance was least. They pushed into a stiff breeze, each of them pointing a beak to the left of their destination while looking to the right of their heading as they crabbed sideways into the crosswind. Stronger migrants—formations of ducks or swans—cruised on parallel courses higher across sky. At eye level, loose groups of small perching birds could be seen in the distance, also flying low over the water.

Ahead, they noticed a few large gulls in orbit above their course, arcing lazily through the air as if there were no wind at all. Gulls were a familiar theme in this landscape. But then one of these birds began to fall, accelerating with strokes too powerful for a gull, into a hurtling dive.

The sanderlings watched as the dark form plummeted, scattering a flock of migrating sparrows. Barely moving its wings, the aggressive bird followed one of the isolated sparrows a few feet above the water while its quarry sprinted to exhaustion. Finally, the raptor stooped low and speared the tiny bird, driving its prey underwater, where its struggles soon ceased. No, these were not gulls but skuas, predatory pirates with the attitude of a falcon and the hooked beak to match.

A second skua pulled into a course parallel with the sanderlings, twenty feet above. Instead of hurrying their pace, the flock tightened and slowed in response. Fifty birds now fit into a space not much broader than the wingspan of the dark form lowering from above.

When the skua moved closer, the sanderlings veered away like a single fluid creature. They would never scatter—their movement was driven by two impulses: approach (closer to each other) and avoidance (of the predator). These common impulses tied their individual motions together. No bird split off from the group, where it would be an easy victim. Instead, the whole flock swerved as one away from danger.

The skua matched their speed. Then it abruptly descended, screaming again and again to frighten the birds into bolting in all directions. A shudder rippled through the flock with each piercing note, but the birds disciplined themselves just as quickly, each pulling next to neighboring birds ahead and beside. Then, before they could react, the skua crashed directly into their midst.

The whole flock dipped down just inches from the water, and still the hawk-sized raptor plunged among them. The big bird had let his vision become immersed in the shifting, turning matrix of rising and falling prey. Working to pick out a unique target from the cluster of identical sanderlings, he had momentarily lost track of one of his own cues—the distance to the surface.

His field of view had been consumed with the motion of individual birds, and he had not perceived that the whole flock had lowered as one before him. So, at the final stage of his attack, he had accelerated headlong into the waves at striking speed, crashing into the water, feet in the air, his wings slapping together above his head.

The sanderlings finally did scatter, abruptly parting before him at the last instant. They bounced off the surface, tumbled against each other, wings slapping the water, feet running along over the waves but never breaking stride.

Just as quickly the flock re-formed its tight defensive ranks, like water filling in over a splash on the sand. The group regained its heading, leaving the skua sitting alone on the waves, facing backward. The big bird dipped up a billful of water and smoothed his ruffled feathers while he glanced at the empty surface around him.

Now the sanderling sat on her empty nest, looking down a shallow tundra slope toward the becalmed inlets and channels of northern Nunavut. Usually this polar landscape was fused into a snow veldt of scattered, low mountains, but for summer the valleys between them melted into reaches of open seawater. The province resolved into a network of narrow straits that carried the tides of the Arctic Ocean beside barren islands lying close across the water to one another.

The sanderling had picked a perfect nesting spot and had been successful. Like other older birds, she had learned the nuances of nesting site selection by trial and error. She had become a master, choosing a site beside the low centerline of a sloping draw, among the strewn rocks. Her motion-less incubating posture had disappeared into the clutter there, beneath her stippled, tundra brown plumage. It was a spot just beyond the track scanned by the eyes of predators.

Her first nesting spot, chosen long ago in her first year, had seemed almost perfect as well—a broad depression nestled amid patches of low plant growth. With her head held still, she had watched white gulls and arctic foxes scan the area as they passed, their gaze sweeping cleanly past her every time. It had been a successful site—until it rained.

She had thought she had a thorough acquaintance with rain, from her travels across many a weather-beaten shore lower on the continent, but the rain that fell in this rare tundra storm behaved differently. The scattered showers had not been torrential, but the thin sheets of water draining from the slopes had accumulated around her, rising steadily. Eventually the water had risen over her nest and eggs, and she was forced to stand. She was uneasy about her own movements so close to the nest—movement attracts predators. But the water standing around her did not recede. Unlike the rainfall she had previously experienced, this layer of water refused to disappear into the ground. This ground was frozen a few inches below the surface, and the frozen layer was impermeable.

The temperature of her eggs fell drastically. For a while they could withstand the change, but they could not withstand immersion—they would suffocate. The moments passed, the waters did not recede, and the urgency of the situation only continued to build. Finally the sanderling acted.

She reached below the surface and picked up an egg in her beak. Then she walked with her head held level, eyes straining downward to pick a path through the low mosses and rocks, and carried her precious cargo up the shallow hill and over the rim. On the dry outward slope, she secreted the egg in a clump of lichens, then turned to retrieve the second of three.

As she approached the low crest carrying another egg, movement ahead froze her in her tracks. Looking over the ridge, she saw the frightening specter of two great black jaegers—oceangoing raiders with long, sharp wings and forked tails—looking down on her transplanted egg. With a stroke, one of them seized the white prize just ahead of the other. Then both birds—one holding her egg—turned to stare in her direction. The second one came after her in a flurry of angry squawks and beating wings. She dropped her egg and flew off as fast and low and far across the plain as she could, never looking back.

Now, years later as a mature parent, she had been more successful. She had built a nest and laid eggs, then left them for her mate to incubate and raise the chicks, freeing her to build a second nest in another location and raise three more chicks on her own. All this was possible because of the seasonal bounty of aquatic insects on the marshy flats, and hard-bodied crustaceans on the seashore. They were far denser than summer's swarms of forage at lower latitudes—in the south, her prey species did not have their whole growing season compressed into a few frantic, snow-free weeks of twenty-four-hour light. Here, on the longest days, swarms of midges rose over the flats in black clouds so high they could be seen a mile away, swaying with the breeze like smoke from a grass fire. Hatchling sanderling chicks grew quickly over the weeks and soon were as big as their parents.

With the lengthening of the summer, the insects and crustaceans were finally thinning out. As the chill of evening approached, the drone of passing crane flies and mosquitoes decreased to silence. The sun had set behind clouds low in the northwest, throwing a purple cast over the islands that stood offshore. Channels between them stayed bright for a while, silvered in reflected dusk light.

The sky would not blacken completely as this night advanced—a thin glimmer would still burn along the northern horizon at midnight—but the evening dimmed enough for the moon to have its effect. It rose full, over the spine of the largest island far to the southeast, sending an avenue of light inward across the waters of the channel—all the way to the rocky beach below. As the twilight advanced, the glistening highway across the water strengthened, darkening the surfaces on either side. Profiles of islands reaching into the flood of light stood blacker still in silhouette.

The sanderling stood quietly, keeping watch as she had all nesting season; she was ready to call out should a snowy owl appear to prey on the juveniles, now foraging on their own somewhere in the darkness. She gazed out over the evolving evening, her eyes drawn to the avenue of moonlight that grew wider as it approached over the water.

And as she stared, the path shining across the sea changed before her eyes, not moving yet growing closer to her—intimately closer. The column of light retreating to the horizon now rose across that distant line to fill the air space directly below the moon. A pillar of light expanded up through space to touch the bright orb and continued up beyond it. The lustrous spire materialized farther above, brightening as it climbed. The sanderling noticed that the path of light on the water no longer stopped at the shoreline but now cut across it, extending over the ground toward her from the beach below—moving up the slope toward her feet.

An involuntary peep escaped her throat. In the blackness beside the tower of moonlight, sparkles were appearing before her eyes in midair. Diamond dust. The distant peep of another sanderling somewhere in the darkness answered her call. The air had grown so chill that a frozen fog was forming all around her—a mist of crystals had materialized in the sky. Isolated flakes nearby flashed moonlight as they tumbled in the foreground. The countless settling particles farther off in the direction of the moon refracted the light above and below it into the radiant column that now bisected her vista. This night was going to grow very cold and frosty.

The sanderling took a few steps up to the top of a rock. The lunar pillar stood out brightly—a portent—its shaft cutting across everything in the background. Moon dogs hung beside it, suspended on either side of the moon. With a flurry of peeps the sanderling jumped into the air and coasted down the slope to the water's edge.

Other sanderlings were there, orbiting the pebbled shore in groups of two or three, landing on the run at flying speed and sprinting to a stop, the cluster of excited peeps coalescing. The juveniles had not come—they would not leave their feeding grounds until there was no prey left. But the rest of the flock stood around, prepared to take wing should the impetus arise. They watched and waited as a final few birds joined the group. What would be the consensus direction if they took flight?

The stones were cold beneath her feet. The sanderling startled as everyone called at once and she found herself airborne, adjusting her direction to avoid crashing into the others all around her. It was too dark to fly in their usual precision formation—the loose group improvised just to fly together straight and steady. They were not bending their course around, as they did on their customary group sunset flight before the night's roost. The sanderling soon realized, as did all her peers, that they were leaving the nesting grounds behind, setting off across the water, cutting through the sparkling air straight down the highway to the moon.

Across the channel their flight rose and fell close over a moonlit plain divided by acres of low, diagonal mounds. They passed through invisible rivers of air, some bracingly cold, flowing from rocky northern exposures, others warmer, coming off water. Unfocused patches of gray smoothed the terrain below them—fog was welling up out of low spots and spreading out over the flats; soon it hid everything at ground level under a blanket only a few feet thick. A band of stratus clouds low on the horizon rose up before the moon, plunging the night into greater darkness. At the same time, a broken skein of ice crystal cirrus far above began to obscure the stars.

So the sanderlings focused on other navigational landmarks—pillars of energy arching so high they disappeared into the sky. These monumental walls of magnetic force rose out of the ground, hinting at the vast metallic foundations of the Earth itself. The birds knew the contours of these stable landmarks, having often relied on them as guides.

And as the flock flew, those soaring magnetic arches came into plain sight. Earth's magnetic lines of force became outlined in green—made manifest by the glistening curtain of the aurora. The sanderling watched them materialize—shafts of fluorescence cascading from above.

Luminescent streamers ignited against a backdrop of red fluorescence higher in the night, setting the clouds aglow with their pastels, bright enough to blind the stars. The sanderling could easily see the arctic plain stretching out in the reflected light. Passing pools of phosphorescence broke the blackness directly beneath her, where lakes and marshlands reflected the atmospheric display. She watched the silhouette of her own flock—black wings beating against a high green backdrop—marking time with her, reflected on the glowing water surfaces close below.

The aurora grew so bright that it became disorienting. The shimmering veils began to ripple along their lower margins, their electricity firing the ground fog into brilliance. The curtains of light swayed, bending away, fusing, twisting. The flock began to twitter, birds crowding each other as the course of their swarm began to waver. Their magnetic landmarks—always constant, in their experience—began to falter.

A jostling disorganization spread through the loose group, bending its course back and forth. Suddenly the flock split, one half banking away sharply to the north. The remaining birds wavered with indecision as to whether to break ranks and fly after the departing group or maintain formation, close beside their nearest neighbors.

The magnetic storm above increased in intensity, a torrent of solar particles raining down, warping the earth's magnetic field. Magnetic north deserted the compass, and the birds' magnetic navigational cues dissolved in the garish green fog. The sanderling did not know whom to follow. She was sure her half-size flock was headed off track as it zigged and zagged— riven by confusion. As they fought waves of reflexive impulse to bank right or press to the left, they dove straight into a wall of thick fog.

The birds around her inexplicably veered around and flew off the wrong way—back behind her, leaving her momentarily solo. She was torn between flying southeast or regrouping with her disoriented band. The cirrus layer thickened overhead, still backlit in a pulsating auroral glow; the high clouds blended smoothly in the distance with the rising ground fog. She realized that she had lost her horizon in the haze and dark.

She called out and heard single members her flock calling back through the darkness, all seemingly flying random courses. Some passed by so close as to risk head-on collision, but when she wheeled about, she could not catch them, their calls fading to the right or left.

Then the sanderling was seized with the fear that she was not flying level. There was no sharp distinction of up and down in the passing fogs. She realized that she might be on course to fly headlong into the rocks below, and that she might do so at any second.

So she threw her wings above her head and fluttered, holding her feet out, looking down through the formless gray pall. What was down there? Deep water? Deep brush? A deep chasm? She continued to parachute through the sky until her wings grew tired of fluttering above her head. How high was she? The fluorescent swirls parted just above ground level, and she finally saw the tilted earth the instant before she touched it.

In contrast to the continuous headwind of flight, the damp soil was dead silent. She bobbed her head once, peeped once, then fell quiet—a formless outline in the dark. Distant calls of single, disoriented birds flying in circles far above grew fainter, then disappeared into the night.

Finally she surrendered to the all-encompassing shroud, folding her legs to kneel against a small rock. The last faint calls she heard were the cries of a flock of geese filtering down from a thousand feet. They talked nonstop, urging each other on, flying exactly the wrong way, back to the north.

Was it morning? A muted dawn had quietly supplanted the wavering glow of the northern lights. The sanderling startled awake, still sitting on the sand, and looked around through the sparse cover. Visibility was less than ten feet; the fog trailed just before her eyes in currents of countless tiny droplets of mist.

The soft silence was punctuated by a single, quick snap—from what direction, at what distance, she could not tell. But she realized that a similar sound had wakened her moments before. She came alert and stood. The sound repeated, perhaps closer—a thin stem breaking.

The sanderling did not know what was out there, but she did know which way was up now, so her decision was quick. With a burst of peeps she jumped up and disappeared into the whiteout.

After less than a minute of spiraling upward, she broke out above a sea of fog, under ropy layers of stratus clouds. The brightness in the east was obvious, so she put the sun direction on her left and set off.

There was a breeze at this altitude. The low clouds below were flowing crosswise over the plain, the strata above moving more slowly northward. As she fell into her flying rhythm, silvered edges appeared around the clouds that hid the sun.

Ahead, the dark, tundra brown ground showed through holes in the fog. When the holes began to merge and the landscape came clear, she began to feel exposed, flying high and alone, so she dropped down to only a few feet above the rocky flats. The tundra scrolled past her—an empty, silent steppe—showing its starkness more clearly when not seen through a wall of other birds. She flew level with the crowns of the struggling plants, skirting shallow, ice-water pools, concealing her silhouette as best she could—this was gyrfalcon country.

The solitary sanderling noticed a flickering, formless motion far off to the left—a shifting cloud of spots tracking along with her, fixed against the horizon. It was a small flock of something. She bent her course to get a better look. Two phalaropes and two dunlins were flying together on a course parallel to hers. She pulled beside, slowing to fly with them. She could not fly close enough to join their slipstreams into a common flock—they flew with different wingbeat frequencies and amplitudes than she. But the company was preferable to the exposure of flying alone.

After a while, she found the birds next to her intruding on her space, pushing her to the right, and she realized they were veering toward another loose formation, flying in parallel with them. It was another mixed flock, mostly plovers. A bigger flock offers more security, so they all joined together. As she found a place in the formation, she saw a lone dowitcher flying in the group, and, to her relief, a pair of sanderlings.

Mixed-flock flying was strenuous—the various birds had different cruise speeds and took up different volumes of space. The slipstreams of air between them were too turbulent to work out drafting distances. The three sanderlings joined up on the underside of the loose formation, keeping their positions as the group made a gradual turn and accelerated slightly toward a formless gray mass floating directly before them.

As they looked through the birds in front, the sanderlings could see they were heading toward the biggest flock they had ever known. It was a disorganized, round cloud made up of every kind of shorebird, all come together after being separated from their own flights the night before. Every migrant along the Atlantic Flyway had been waylaid by the freak conditions.

Her smaller group fused with the ponderous assemblage, plunging into an unwieldy mix of constant squawks and peeps, amid a rush of wings that sounded like surf. Yellowlegs, turnstones, surf birds, whimbrels—birds of all different sizes were flying slower than they would like, working to adapt their styles to the others! Individuals were not watching where they were going but were looking sideways, hoping to find members of their own species. There was constant jostling as groups of three or four sandpipers or knots cut in front of everyone else to join up beside others of their kind, with whom they could more easily fly in formation.

The chaotic throng was sorting itself into subflocks. The sanderlings were coming together with other similar-sized sandpipers, their compatible aerodynamic profiles providing them a comfort zone next to the bigger birds. The flock continued to balloon as solitary migrants raced across the tundra to join the security of the group. Newcomers kept the confusion level high as they made their way through the crowd in search of their own kind.

Then the birds around the sanderling grew quiet, and the whole group contracted. Each member peered through the flurry of wings and caught sight of the gyrfalcons, flying in parallel to the right and left. This high, slow flock would inevitably attract attention—the swarm was visible for miles.

The predators were themselves deterred, keeping their distance, having never seen such a flock. They were now only silhouettes holding their positions against the low horizon, but there were lots of them, solitary hunters drawn into a pack by the conspicuous aggregation.

Each time the sanderling looked off to the right, then back again left, the falcons appeared higher and larger. The bird hawks balanced the intimidation they felt with the reassurance they found in their own uncommonly high numbers, and soon they were emboldened to close up with the margins of the extraordinary migration. Their innate aggression was tempered with curiosity—this swarm of birds before them was too massive and unwieldy to respond with the expected evasion; its members pressed straight ahead, leaving the falcons looking at each other. Finally, one of them began to call out, a harsh, grating cry that was picked up by the others. All around the periphery of the massive flock, the falcons set up a barrage of cold taunts.

With each piercing note, a wave of indecision cleaved the crowd of birds. Every one tried to pull deeper into its own embedded flock, no longer mindful of the different species all around. Claustrophobia pressed against each individual as the walls of other subflocks blocked their impulses to break right or left; slower groups ahead or above constrained their maneuverability.

The pressure was released in the blink of an eye. As the cries of the raptors escalated, the loosening shoal of birds suddenly exploded, flying apart into all its smaller, homogeneous subflocks. They careened away from one another, now released to bank sharply or dive away—leaving the space between them free for the last birds of each kind to sprint through open air and catch up with their own.

The falcons chased after the hindmost, their pursuit response triggered by the sudden panic. They seemed drawn to the larger of the trailing birds, and they cut between flocks of the smaller ones, rolling past startlingly close in front of the sanderlings.

As her flock pitched hard in the opposite direction, the sanderling was stunned by a crushing impact from above that killed her forward motion. As she began falling, her feet closed on something and she was dragged through the sky while she came back to her senses. She was clinging to feathers, to the nape of the neck of a raptor fiftyfold more massive than she. She lay on a moving platform—wide, white wings pumping through the air but failing to gain headway as they thumped with each stroke against the random impacts of small sanderlings, each sent spinning sideways.

The monster screamed, deafeningly close, and the sanderling loosed her grip and tumbled into the slipstream of the falcon's passing. In the same motion, she rolled free, found her horizon, and bent her course toward the nearest sanderling—which was bending its own course toward two more sanderlings, all peeping excitedly.

She sensed a center to the sound of all the scattered calls, as did the others around her, and, in mere seconds, their shattered flock had reorganized itself. They found their bearings and dove for the ground together, heading southeast.

The sanderling never looked back. She gave no thought to the success of the falcons, or to whether they were trailing her. She was focused on her own bruises and ruffled feathers, on her breathing, and on holding her position within the formation. She could tell by the ambient sound that her mixed group of sanderlings and sandpipers was now alone. The other flocks had scattered in all directions—the sky above her flight was empty.

The birds around her had fallen into their favored rhythm, flying just as they would all that night and the next day, until they came to the North Atlantic shore. Before dawn, they would be met by Sirius—a bright, blue-white harbinger of autumn to guide their flight as it climb high before them. It would remind them of Canopus, a similar beacon that would rise a few months from now to oversee their second summer far to the south.

The sandpipers seemed to want to fly more southerly than the sanderlings did; the other birds accumulated on the south side of the flock. Eventually a canyon opened in the midst of the group, the wind of headway spilling into the sheltered center of the flock as the two subflocks separated, her half pulling away on a more easterly heading. This left the sanderlings to come back to the course they knew best for this leg of their journey. They flew easily, secure with other sanderlings on every side, watching the afternoon sun marking time with their passage, the distant low mountains gradually passing beneath it off to the right.

Science Notes

Sanderlings follow varied routes, migrating up and down both North American seaboards annually; the flock here pursues an annual clockwise migration all the way around North America (Macwhirter et al., 2002; Myers et al., 1990). They experience the broad, contrasting range of North and South American shoreline biogeographies, including the attendant oceangoing and terrestrial predators. Their migration keeps them in fair weather—on the summer side of the equator. In this tale, they engage in the less common behaviors of mixed flocking and egg carrying.

Avian migrants follow a course guided by the light polarization in the sky (Muheim et al., 2006), as well as by geographic, stellar, and magnetic navigational cues. Cells within birds' brains detect the Earth's magnetism, and it has been proposed that through a linkage between those cells and the visual cortex in their brains, they actually see magnetic north superimposed in their visual field (Heyers et al., 2007). In this tale the sanderlings' magnetic compass navigation through the subarctic is thrown off course when a magnetic storm disrupts the Earth's magnetic field—an auroral bombardment of the atmosphere by charged particles from the sun.

References

Heyers, D., et al. 2007. A visual pathway links brain structures active during magnetic compass orientation in migrating birds. PLoS ONE 2: e973

Macwhirter, B., et al. 2002. Sanderling (*Calidris alba*). In A. Poole and F. Gill, eds., *The birds of North America*, no. 653. Philadelphia: Academy of Natural Sciences; Washington, D.C.: American Ornithologists' Union.

Muheim, R., et al. 2006. Polarized light cues underlie compass calibration in migratory songbirds. *Science* 313:837–39.

Myers J. P., et al. 1990. Migration routes of New World sanderlings (*Calidris alba*). *Auk* 107:172–80.

Unseen Masters of the Sea

The shark cruised steadily, just below the surface. She could see individual stars wavering through the water and kept them in place to the left and right, guiding her navigation to the west. She was so strong and so long— and yet so sleek that the laziest of ripples back and forth along her spine sped her onward at a pace she maintained day and night.

She was a constant cruiser—a fish that would not cease her forward progress until the day she died. When she broke away from her curving search patterns above the reefs of the shallow water and headed straight out to sea, she would migrate nonstop, commuting between continents in a matter of weeks.

She rode at the peak of a broad food pyramid comprised mostly of lives too small even for her eyes to focus on as they slid past. Already she was bigger than 99 percent of the other creatures in the ocean. She had survived and grown by following her basic drive to attack small moving objects and to flee from larger ones. But she also had been born with the capacity to adapt to a changing ocean, and the ocean was indeed changing around her. The web of life on which she depended had fallen under the control of another group of animals—nebulous, transparent forms that had recently come to dominate the world she lived in.

Her track across the Indian Ocean had taken her into a region where those vitreous jelly creatures began to appear in the water, first far off to the side but soon more directly ahead. The ornate, dimly fluorescent globes were increasing around her; glassine threads below the closer bells were barely visible where they streamed away into the depths.

In the distance she sensed the fin noise of a disabled fish—the uncoordinated spasms of an animal no longer able to stay vertical in the water. Because of prowlers like the shark, such sick or injured fish rarely suffered long before they died. Her course automatically bent toward the sound, accelerating as she sought out her target.

She attacked as soon as she located the stricken animal and immediately felt the prick of barbs through the thick skin on her gills and snout— she had strayed into the curtain of poisonous tentacles trailing unseen from one of the jelly creatures. A sharp pain on her left eye caused her to spit out the bleeding fish and break sharply away. She abandoned the surface and dove to two hundred feet to continue on through blackness below the reach of the trawling strings of nettles.

At this depth she relied on navigational cues felt but not seen. She could sense the slope of the Earth's magnetic field slanting through the space around her, and she held her compass fixed at an angle to that slope. She marked the passing gradients of temperature and flavor. Scents here were subtly altered by the pressure; sounds carried with much greater clarity and detail at these depths. But the creatures that inhabited this space concealed their scents and sounds—and their forms—in the darkness.

With no forewarning, the shark collided with a soft wall in midwater. A sheet of mucous filled her mouth and she shuddered to a stop, choking. She whipped her snout back and forth, and immediately a string of lanterns lit off to one side, moving forward in unison with the swinging of her head. As she fought to free herself from the entanglements, other flares ignited until she was lit from all sides. Rows of bright lights drew together around her as if linked, dispelling the blackness and binding her like an anchor cable winding around a winch.

She had blundered into a fluorescent siphonophore, and now every move she made tangled her more deeply in its linear body. The mild stings of this jelly creature festooned over her fins grew increasingly obvious. The siphonophore was a predator not of large fish but of shrimp and other crustaceans. It was trying to disengage itself as well, retracting its web of tentacles, but its efforts to pull away were thwarted by the ever more desperate movements of the shark. When she could not move forward, the shark was on the verge of panic.

Sculling her broad tail backwards, the shark measured the sinuous form of the creature she had encountered. She focused on a dark center in the ring of lights and bolted for it. Again she met the soft wall, but she surged ahead and the resistance yielded.

From the point at which the gelatinous animal was ripped apart by the shark's impact, the string of lights started to grow. One by one the line of phosphors extended to reveal the entire length of the broken being. The shark was lit from head to tail by the glare from chains of lanterns ten times her own length. One chain filled all the water directly above her like a ceiling; a similar coil contracted on itself to her right and below. The two sundered strings glided apart like floats on lost lengths of driftnet. She broke away to the left and dove for even deeper water.

The siphonophores share a strategy with other select members of the dominant phyla in the sea—the jelly creatures. All of them pursue their adult lives without benefit of hard parts—they let the water support their weight. This strategy has left them as flexible, passive floaters, spiders in webs of tentacles, waiting for the active swimmers of the realm to blunder into their snares. Were success measured by the weight of all the individual jelly creatures, theirs could be the most successful predatory strategy in the ocean.

Their fishing line is flexible, extending when their startled quarry bolts at its touch, later retracting when the victim's flight has been reduced to futile spasms. The venom in their sting makes up in its potency for the absence of active pursuit by the jelly creatures. The less venomous of their numbers, such as the ctenophores and the pyrosomes, simply absorb planktonic prey onto their sticky surfaces.

Freedom from the demands of a skeleton has removed limits on the size of these creatures. They attain some of the largest dimensions in the sea—siphonophores have been seen stretching to nearly one hundred feet. The lion's mane jellyfish has tentacles that reach farther than the length of a blue whale; its flexible filaments trawl for prey far beyond the warning silhouette of the shaggy float.

We have very little knowledge of the final dimensions of these creatures or of the impact they have within their realm—they are as difficult to study as they are transparent. Analysis of trawl samples of drifting species allows the census of hard-bodied animals, but such studies destroy the gelatinous fauna, reporting them only as "goo" fouling the netting. The midoceanic forms grow the largest, but they are the most fragile because they do not have to tolerate the wave action of coastal waters. The bigger they are, the more likely they are to be shattered by our attempts to retrieve them from the depths and measure their numbers and their size.

The shark cruised along an Indonesian beach, sampling the myriad suggestive flavors that swirled beneath the shore break. A school of coastal shrimp parted before her, the individuals swimming in start-and-stop pulses. The clicking of all their tails increased to a buzz as her approach moved them en masse from her path.

Thundershowers had passed. The sky was clearing, the surface calm above a rising tide. A flood of cold freshwater poured from the mouth of a canyon in the headlands up from the shoreline. The outflow spread across the surface, carrying with it fragrances washed from a land known to the shark only from scents carried downstream to the sea in rainwater.

She was nearly mature now and had learned to exploit all the various hunting grounds the seas held, including the near-shore environment. She did not usually eat things from above the waterline, but she was curious and impulsive and always hungry. As she turned back toward the sea from a mangrove delta, a gossamer blue box coming in from the open ocean deflected her course. Immediately she perceived several more of the creatures, all pulsing toward the beach, forerunners of a swarm of sea wasps.

The shark was well acquainted with swarms of jelly creatures. She had once encountered a shoal of greenish floating goblets so extensive that she had swum beside it all day long and never come to an end. Eventually she gave up the attempt to go around and dove to resume her course. Then she watched green lights passing above for more hours still, like lamps hung from an infinite ceiling.

But these sea wasps were some of the most virulent of the jelly creatures she had encountered. One of them had once wrapped a tentacle across her flank. The fiery sting had driven her to bolt sidelong into the sand to scrape away the torment. It had taken her days to recover from the sickness that followed.

Now she faced a wall of these creatures increasing in numbers, all moving in on her from the open ocean. They paralleled the shore as far up and down the beach as she could see. These box jellyfish were swimming rapidly, following the freshwater scent on the surface. The shark moved aside along the shoreline, but the sandy slope soon curved seaward, pushing her course toward the oncoming tide of jelly creatures. She was trapped.

So she settled to the bottom, once again facing the panic that arose when she could not move forward, yet immobilized by the memory of the stinging tentacles now trailing toward her again along the seafloor behind the advancing front. The sea wasps were mounting a coordinated assault on the island, closing in on the scent of the freshwater that poured downstream into the ocean. They would be invading the beaches, swimming up the inlets, their lethal dragnet trapping prey that lived along the freshwater boundary or was carried in the runoff.

Resting on the slope of a sandy rise, the shark watched the armada of pulsing ghosts closing overhead. A pair of green turtles sat impassively on the coral nearby, each eating a box jelly with no ill effects; live severed tentacles drifted off on the current. She knew that if she moved, these barely visible stingers would stab her, but they would soon descend on her from above even if she stayed in place. She could not remain motionless for long—her life depended on her forward motion.

She fought the urge to bolt through the curtain of barbs—an excruciating gauntlet she would not likely survive. She rose to hover above the sand, only to be startled by the appearance from nowhere of a large shape right in front of her, flashing close before her nose and diving to the left, behind the rise. A long-dormant instinct for schooling spurred her to follow, and she leapt from a cloud of silt, hugging the bottom as she turned sharply over the rise to follow the leader.

Straining to match the other's speed, she followed a path that retreated toward the shore but descended slightly nonetheless—a channel that provided some separation from the swarms of wasps on the surface. Like her, the leading animal was escaping the jelly creatures, sending waves of pressure through the water with the momentum of each stroke of its narrow, flat tail. She realized that this beast was larger than she was, but it was not moving as fast as its bulk might suggest. Though it was covered with large scales, it did not swim with the grace of a fish.

She caught a better glimpse along the creature's length; its eyes were not on the sides of its head but on top. This was a land creature—a crocodile— its feet folded back along its flanks. It now showed none of the aggression that the teeth poking up and down through its closed jaws implied; it was retreating purposefully toward a solid rocky wall. The obstacle would shortly bend their paired courses up into the surface now crowded with ranks of box jellyfish.

The crocodile did not slow or rise but pressed on toward the base of the looming rampart toward a point that grew lighter as they approached. A tunnel was opening before them, an archway in the rocks that the crocodile entered. The shark followed. Their speed carried them swiftly through the passageway, delivering them to deeper water where it opened on the far side. They had swum under a rocky headland, beyond which the beach curved away from the point. Following that curve, the two fugitives distanced themselves from the armada of box jellyfish, which was moving in the other direction here, back up the shore.

The crocodile rose through the water toward the beach. It slowed when its head merged with the surface, and its feet spread open to meet the sand. There it continued its progress on foot, disappearing through the waterline, its tail the last to slip into the mirror ceiling. The shark continued at depth and returned to open water.

The open ocean is both the widest and least known habitat on Earth. It covers 70 percent of the planet to an average depth of more than two miles—the measure of a world largely unknown to us. The jelly creatures ride invisible rivers of current through this void. They are as intangible as the conditions that govern their numbers—levels of light, degrees of water temperature, upwelling tides that foster the multiplication of their prey.

They occur at all depths, beginning at the surface, where schools of medusae pulse their bell-shaped bodies in seemingly random directions. Those jellyfish are members of the same phylum as the anemones and the corals. Other jelly creatures include the salps and the pyrosomes—members of the same class as the vertebrates, though as adults they have resorbed the backbones that were present in larval stages. The armored phyla are also represented here by the sea butterflies—a family of mollusks without shells that float on transparent oval wings.

Jelly creatures materialize in countless millions at confluences of favorable circumstances. They often include an asexual phase in their reproduction cycle, allowing their numbers to respond quickly to productive conditions. They can rise in a matter of days to become the predominant creatures by mass in any given location. They have commanded a position of dominance for hundreds of millions of years—for as long as there have been multicelled life forms in the ocean.

The jelly creatures first appeared six hundred million years ago, predating the fish, and have flourished while other taxa have come and gone. Since they preserve poorly as fossils in the shale, our assessment of their past prevalence will always be an underestimate, as is our evaluation of their current impact. They multiply like weeds through an oceanic environment that is going through a period of disturbance; they flourish in polluted harbors and river deltas. They appear to be increasing now in the Gulf of Mexico and the Western Pacific, plugging intakes and shutting down shoreline waterworks with the masses of their bodies.

No matter how far across the seven seas the shark traveled, the gelatinous fauna seemed to grow thicker. This was just as true, she saw, in the Mediterranean. Here, the creaks and groans of fish and whales had been replaced by clanging and surging sounds transmitted into the water through the shadowed, featureless hulls of ships plying the waves above. The shark swam beneath those floating hulks, following currents tainted by foul scents and animated by few swimming creatures. She was heading back to an inland sea she had once visited, a place where there had been fewer ships and more fish than in the marine desert she now traversed.

The fishing boats had been taking the larger animals out of the water. With less competition from the fish, the jelly creatures found more drifting crustaceans—amphipods, copepods, shrimp—to eat. So the jellies increased in number and became significant predators not only on planktonic prey but also on fingerling fish and the floating eggs from which they hatched, depressing the recovery of the stocks of larger fish even further.

As she navigated below the hulls of the fishing boats, the shark noticed pieces of dead meat suspended around her. She did not take this bait because she had learned to see, in the shafts of sun, the glint of the fishing line attached to it. The line looked to her like strands descending through the water behind jellyfish—holding immobilized victims, likely to carry stinging barbs.

She finally left the Mediterranean to swim against the current flowing west through the Dardanelles. There the shark found the waters filling with small jelly creatures riding the tide the other way. These were ctenophores—comb jellyfish only a few inches long, surrounding her in growing numbers. The pellucid creatures moved at random, driven by girdles of iridescent cilia that chased dashes of rainbow color around their oval bodies. Their numbers multiplied into a blizzard of suspended spheres slipping horizontally backward beside her as she pressed ahead.

The current slackened when she entered the Black Sea basin, but there were no fish to be caught there. Ctenophores glided aimlessly everywhere she looked. After days of searching she was ready to eat a decomposing gull off the bottom if she could find one, and the crabs around it as well. But there was nothing at all to be found but milling ctenophores mile after mile after hungry mile.

In the early 1980s ctenophores were introduced to the Black Sea—apparently by the carelessness of humans—into a habitat that had never known them. They were carried from the waters of North America, where late summer peaks in their population were usually followed by spikes in the population of their predators. In the absence of such checks on their numbers, the ctenophore population exploded. By the summer of 1989 they had decimated the Black Sea's plankton, severing the food chain at its first link. The fisheries on all Black Sea shores crashed, never to recover, while ctenophore numbers surged, partly at the expense of the remaining floating fish eggs. By the 1990s the ctenophores had found their way through an artificial canal into the Caspian Sea, with the same consequences.

In the late 1990s, apparently by another spin of the wheel of environmental roulette, a second alien species of ctenophore was introduced into the Black Sea—a species whose niche is that of a predator on the first species. The second species then experienced its own population explosion, and by the turn of the millennium, the densities of both comb jellies in the Black and Caspian Seas were coming into balance with each other.

We cannot say with certainty that the populations of inedible jelly creatures are growing around the world as a consequence of our exploitation of the oceans—in response to growing water pollution and unsustainable fisheries. Nor can we predict how much these creatures will continue to increase at the expense of the remaining fish eggs and fry. Blooms of jelly creatures that occasionally cover thousands of square miles of ocean surface may merely be normal variations in their population cycles. Our fuller understanding of their role in the changing ocean awaits our deeper commitment to an appreciation of the biology of the sea around us, and then a fuller accounting of our impact on it.

Science Notes

The primary consumers of the photosynthetic algal plankton are the first link in the oceanic food chain. These consumers include the larval stages of all the major oceanic phyla—the fish, the mollusks, the arthropods, the echinoderms—as well as zooplanktonic predators that are still microscopic as adults. At the top of this food chain, in competition with the warm bloods—the birds, the whales, and the semi-warm-blooded top predator fish (the sharks and tunas)—ride the jelly creatures. They are one of the oldest (Cartwright et al., 2007) and most successful groups of creatures in the ocean. They make up, by mass, the most common form of macroscopic life on Earth, yet they are relatively unknown to us. They go by many names not commonly heard: hydromedusae, scyphomedusae, pteropods, polynyas, heteropods, doliolids, pyrosomes, larvaceans, siphonophores. Most of them go through an asexual phase in their reproduction that allows them to multiply rapidly when conditions turn favorable. These are self-fertile hermaphrodites, and their numbers increase immediately in response to increases in plankton. Plankton can increase due to increased nutrients washing into the ocean as pollutants from rivers and shores; they also increase with the decrease in predatory pressure upon them, when their finned predators are fished out.

Of the primary oceanic phyla, many of those known to us by their hard forms are also represented by jelly creatures (see table).

Glossary of Some Jelly Creature Names
Ctenophores were erroneously classified as jellyfish until it was discovered that their stinging cells are not their own but are transported intact to their surfaces from the tissues of the jellyfish they eat.
Cubozoans are box jellyfish. Active swimmers with eyes hanging vertically from their cuboid bells, their vision guides them past obstacles in shallow water. Their most poisonous member (*Chironex fleckeri*) preys on animals of freshwater stream mouths and estuaries. It is a fragile creature but well protected by its sting. One of the most venomous creatures in the sea, its poison is quickly fatal to all that contact it, including humans.
Pyrosomes are colonial pelagic tunicates; found in tropical waters, they grow up to four meters in length; they move water with cilia, exhibit diurnal migration, and are brightly bioluminescent when disturbed—visible for hundreds of yards through the water.

Salps are free-swimming tunicates in which the spine is present only at the larval stage. Adults have resorbed their bones; they pulse their body wall to move water and propel themselves.

Sea butterflies are free-swimming mollusks without shells.

Siphonophores are colonial jelly creatures, such as the Portuguese Man o' War; most of them inhabit the deep sea.

Tunicates are (usually sessile, or settled on the bottom) chordates that filter feed by pumping water through their hollow bodies.

Oceanic Phyla, Including Those Represented by Hard and Jelly Forms

Jelly Creatures*	Hard Creatures
Mollusks	*Mollusks**
Sea butterflies (Corollans)	Clams, snails, nautaloids
Vertebrates	*Vertebrates*
Salps (tunicates)	Fish*, whales
Cnidarians	*Cnidarians**
Jellyfish (Scyphozoans)	Corals
Box jellies (Cubozoans)	*Arthropods*†
Colonial jellies (Hydrozoans)	Shrimp, crabs, copepods
Ctenophora	*Echinoderms*†
Comb jellies	Starfish, sea urchins

*These have planktonic larval forms.

†These do not include members with a jelly form.

Jellyfish blooms arise through unpredictable confluences of events. The jellies and their prey may be concentrated by winds and currents (Graham et al., 2001). Converging surface currents deflect each other downward where they collide. The efforts of the jelly creatures carried on those currents to press back to the surface result in their accumulation at the convergent boundary. The modern increase in the number of jelly creatures may be due to the effects of overfishing (Mills, 2001). When the fish that compete with them most directly are removed from the top of the food chain, their mass can grow to overtake that of the remaining fish (Lynam et al., 2006).

Lamnid sharks (mako, white, porbeagle, and salmon sharks) are committed to perpetual forward motion. They have no air bladder; if they do not swim, they sink. They do not pump water over their gills; it flows over them as a result of the animals' forward motion. If the flow stops, the animals suffocate. Their swimming muscles are kept warm, like mammalian muscle, by heat generated by their constant motion; lack of motion by the shark causes those muscles to cool and become dysfunctional. When the muscles cool, the shark cannot swim (Bernal et al., 2005). Their commitment to constant cruise mode disposes these sharks to a migratory lifestyle (Bonfil et al., 2005).

Disruptive marine species are introduced into vulnerable habitats when ballast water taken in by ships on one side of the world is dumped in ports on the other. Ballast water is taken on board to stabilize a ship that would otherwise be floating too high in the water because of empty cargo or fuel holds. Environmentally careless disposal of transported ballast water has led to problems on shores the world around. Oil pollution results when ballast water carried in fuel tanks is emptied. Successive ctenophore introductions, presumably from ballast water (Shiganova, 1998; Shiganova et al., 2001), are only part of the story of the abrogation of stewardship responsibility for the waters of the Black Sea—a story that also includes overfishing (Daskalov et al., 2007).

References

Bernal, D., et al. 2005. Mammal-like muscles power swimming in a cold-water shark. *Nature* 437:1349–52.

Bonfil, R., et al. 2005. Transoceanic migration, spatial dynamics, and population linkages of white sharks. *Science* 310:100–103.

Cartwright, P., et al. 2007. Exceptionally preserved jellyfishes from the Middle Cambrian. PLoS ONE 2: e11121

Daskalov, G. M., et al. 2007. Trophic cascades triggered by overfishing reveal possible mechanisms of ecosystem regime shifts. *Proceedings of the National Academy of Sciences, U.S.A.* 104:10518–23.

Graham, W. M., et al. 2001. A physical context for gelatinous zooplankton aggregations: A review. *Hydrobiologia* 451:199–212.

Lynam, C., et al. 2006. Jellyfish overtake fish in a heavily fished ecosystem. *Current Biology* 16:R492–93.

Mills, C. E. 2001. Jellyfish blooms: Are populations increasing globally in response to changing ocean conditions? *Hydrobiologia* 451:55–68.

Shiganova, T. A. 1998. Invasion of the Black Sea by the ctenophore *Mnemiopsis leidyi* and recent changes in community structure. *Fisheries Oceanography* 7:305–10.

Shiganova, T. A., et al. 2001. The new invader *Beroe orata* Mayer 1912 and its effect on the ecosystem in the northeastern Black Sea. *Hydrobiologia* 451:187–97.

Water World

When she first left the island of her birth, the petrel found herself lost on a wild and trackless sea. With no idea how to find her way to her next meal, she set off chasing the other petrels out over the open ocean. They taught her how to survive, and then she taught herself how to prosper—as she learned the subtleties of her world.

Now, after years living on the sea, the petrel had learned that the broad vista was covered with signs—crossroads she could follow much more easily than she could navigate the features of the solid ground of her birth. Solid surfaces had become deserts to her—places of short lines of sight frequented by predatory gulls. Vertigo came to her quickly when she chanced to light on land and found the regular heaving and falling motion of the sea abruptly stilled.

She had matured into a creature of the open ocean. She had not seen land, smelled dust, or heard surf for months. Sky and sea fused uninterrupted along her horizon as always. Her own flock filled the air around her, while groups of dark terns and shearwaters followed their own pathways over the water in the distance.

The longer-winged seabirds, the albatrosses and the boobies, were nowhere to be seen—they had been blown a thousand miles away by the gales of the previous week. The shorter-winged petrels had sliced through those winds, banking close over the crests of the waves to cut through the troughs with one wingtip inches from the water.

Parallel ranks of cumulostratus clouds stretching from one edge of the sky to the other were the last remnants of jet-stream winds that had finally abated. But one cumulus column stood out among the rest. It grew higher than the others, drawing the petrel toward it. That billowing tower was born in updrafts that rose from the flanks of a volcanic island standing below the far horizon.

The petrel knew that the days of gales had increased the strength of the current flowing around the island. The wake the island left in that current was now swirling with great eddies. Those deep whirlpools were stirring up the bottom, like tornadoes moving across the plains—mixing the deep, nutrient-rich silt back up into the surface layer. That levitated plume of nutrients trailed away across the ocean, fostering a bloom of plankton in the island's lee. This broad green swath on the sea would be alive with the predators that thrive in the webs of the plankton-based food chain.

Without ever seeing the island, the petrel caught the scent of the planktonic highway on the breeze and turned upwind. Soon she found the waves animated with flashing silver fins under a cloud of fluttering white wings. The air above the swells was alive with prion petrels—small, filter-feeding birds that ran across the surface on webbed toes. They dipped up billsful of water and seined out the plankton, pursuing the same feeding strategy as their much bigger brethren, the baleen whales.

The petrel climbed above the fray forty, then sixty, feet, more than doubling the distance she could see with every second she rose. She saw the prions gathered by the thousands, but she was looking for something else—her next avenue across the water—the track of dolphins.

Cries arose around her as the other petrels turned and dove to the west. As she wheeled to join her flock, the dolphins appeared ahead in the middle distance, their course across the sea marked by trails of mist suspended above their blowholes. The petrels closed the gap, turning downwind to catch up with the pod from behind; then they slowed to keep pace above it.

The sleek mammals showed in clear detail through the surface. They were coasting, chasing each other over short sprints, keeping a loose school. Like ribbons trailing from their flanks, dark remora fish clung to some of them, releasing their suction hold occasionally and dashing from one dolphin to another. Diminutive newborn dolphins swam in close formation with their mothers.

Because the petrel now coasted, flying the same speed and direction as the wind, the air around her seemed calm. The sigh of the breeze blowing through feathers was stilled in her ears, allowing her to eavesdrop on dolphin vocalizations that carried through the surface. She heard the rapid bursts of dolphin laughter as pairs of them cavorted, and the steady creaks and pops and whistles of conversations elsewhere in the pod.

As she kept pace, gliding above the group, she heard the chirps and trills below become more lively. Then she squawked at a sudden breach just before her and banked sharply away. Dolphins leapt from the water, throwing spray as they spun through the air to land at a different heading. Their pod changed course, and the petrels keeping pace on the other side of the surface turned with them.

From above, the dolphin's tails appeared to pump up and down more vigorously now, rhythmically brightening, then dimming, as the creatures drove themselves faster below the waves. The petrel easily matched their

Walk on the water. A petrel supports its flight just above the surface by skittering across the waves.

speed. Gradually she sensed the school slowing again as other shapes came apparent in the shadowed light farther below. The deeper creatures could have been more dolphins, but they moved differently—swaying their tails from side to side. Those tails appeared dark and thin. The dolphins had joined to swim above a larger and deeper school—a school of tuna.

The moving surface of the water blurred as the petrels kept pace with the paired schools below. Together, they formed a tiered front advancing to the southeast—fish in the dark blue below, birds in the bright sky above, mammals and the sea surface in between. The fleet proceeded at a steady five knots, tacking eastward, eventually coming around to a course toward the morning sun.

Then the progress of the convoy slowed and began to spiral into a tightening curve. By watching the birds of her flock milling about and crying through the air, the petrel saw that the entire group was bending its course around upon itself, soon to close into a wide circle. The seabirds above, the dolphins below the surface, and the tuna below them were all moving in unison in a constant arc—a whirlpool of sleek bodies with its upper edge in the sky and its base lost in the depths.

Eventually, the center of the rotating gyre of dolphins and tuna began to fill with a great, amorphous shadow—its margins flecked with glints of silver. As she wheeled above, looking down across her lowered left wing, the petrel watched the texture of the shadow brighten. Its thousands of members resolved—a cloud of small fish suspended within a moving, curving cylinder of sleek flanks.

The tuna had found a school of herring and were driving them upward. The small fish moved lazily in their parallel ranks, showing their sprints only where individuals spun inward, dashing from contact with the circle of hungry corsairs closing in around them.

Shoals of silver flashed in unison, then dimmed, as the formations of herring sought shelter within their own numbers, only to be herded into a tighter and tighter ball by the tuna and then the dolphins. The smaller fish contracted into a single, reflexive organism, avoiding only the most imminent danger, mindless of the barrier they were about to be trapped against—the ceiling of their ocean.

When the herring finally touched the surface, they had packed themselves densely together. As the school was squeezed further by the press of the closing ring of dolphins, it reached a claustrophobic density that prevented the fish from moving their fins freely. At that point, the herring panicked. The smooth flow lines of their school dissolved into a milling scrum as they dove into each other. The dolphins and tuna moved into the fray, snapping blindly but successfully, then reversing to chase the smaller prey that exploded through the cordon of jaws, separated from the safety of their numbers.

The smaller fish threw themselves through the surface, escaping the teeth of the dolphins only to be set upon by the birds orbiting above. The petrel dove into the fray, foraging across air, liquid, and solid boundaries simultaneously. She stumbled through a noisy fracas of bubbles and foam, constantly pumping wings that drove her alternately through the air, then through the water.

As she pursued a leaping fish, she crashed straight into the face of a wave and then out the other side, flying and swimming with the same motion. With her wings held above her and her feet hanging down, she ran across the water as she chased along the slant of a swell, closing in on another silver target that was snapped up by a dolphin inches ahead of her. Because dolphins are the same shape as the great fish that eat petrels—the marlin, the sharks—the petrel reflexively cut sideways away from the dolphin's fixed smile, running up its flank, and was blown into the sky by the warm geyser from its blow hole.

The birds who had caught their fill rose through the rotating flock to glide along its upper rim, looking down on the feeding frenzy, waiting to get hungry again. Then they spiraled down through the cylinder of wings and slowed themselves by lowering their feet, looking for more target fish. The flock fed on the surface, sometimes diving a few feet under after a selected target but not lingering long in the melee below the waterline.

Feeding continued until the school of herring had dispersed into individuals too scattered to find. The dolphins and tuna dispersed after them, and the birds lost track of the schools and settled on the surface. After the predators had gone, the herring reassembled their community from empty water in minutes, individuals coming together to smaller groups that all soon coalesced. The reconstituted school—noticeably smaller now—gradually sank back into the darkness.

The petrel was at home on the ocean—her adaptation to the ways of the fish superior in many ways to that of the fish themselves. She swam and dove like a fish. She had no more need for dry land than a fish, living out of sight of shore for years, drinking salt water, sitting in the swells as comfortably as a sparrow perches on a branch or a willet stands in the sands.

She could see farther across the surface, and almost as far below the waterline as a fish could see. If she heard an eruption of squid or flying fish throwing themselves into the air to avoid a predator, she too could burst into flight and escape the water. Once clear, however, she was not bound to fall back into danger like a leaping fish but could gain altitude. Nor was she compelled to flee blindly; she could circle back to investigate the threat. Was it only a turtle? A manta? No threat at all. If it was something bigger, sleeker, she would gauge its intentions from a safe height and then take off in the opposite direction, calling out to warn the rest of her flock to do the same.

The seabirds are wonders of reverse evolution, having bettered the fish by taking that giant step out of the water long ago. The feathered colonists of the land had learned to live in the air, to breathe it and fly through it; then they reversed their steps and returned as air breathers to the water world from which they had departed. The medium they inhale is 20 percent free-flowing oxygen, whereas the fish had to push the greater resistance of water over their gills—a medium that is less than 1 percent oxygen. The more ready aeration of their blood gives the birds a much greater aerobic metabolism than that of the fish. The birds have the greater stamina, whereas the fish are short-distance sprinters. Petrels are capable of feats of migration that give them access to widespread or transient resources far beyond the capacity of the fish to exploit.

The petrel was attuned to the sensation of things brushing against her feet as she sat on the surface. Such contact could be startling—one male in her flock had only one foot left, after an encounter with a sea lion. Nonetheless, she usually ignored the caress she felt when kicking the occasional shred of flotsam. But one night long ago, in her first year, she had felt something catch hold of one of her feet, followed quickly by the sting of pain. She leapt from the surface and cried out, sending everyone in her flock airborne. But the grip on her foot did not release, growing heavier the higher she flew, though she could not see anything pulling her down. Finally she saw the dim luminescence of a thin filament trailing back to the water, its end dipping through the surface farther away from her launch point the higher she climbed. When she had finally pulled the strand clear of the water, she was forty feet in the air, and the extra weight was quickly exhausting her wings.

She sank through the sky, fluttering ineffectually as she pecked at the painful area around her webbed toes. She continued to work at the gelatinous tether after she splashed against the water again, and finally it released. It was not attached to anything bigger; it did not chase after her; it did not swim at all. It was merely a detached thread from a jellyfish, drifting passively on the currents. But its stinging cells were still alive, and she would feel the debilitating aftereffects of their venom for days.

Now she bobbed on the surface, a seasoned two-year-old bird, paddling occasionally—just enough to maintain her headway. She would work harder if a rogue breeze materialized from the night to nudge her sideways, but now it was calm. She sat idly under the stars, tasting the evocative scents that passed by just above the surface—traces that grew clear enough to recognize for brief seconds before they faded again beyond her grasp. Some of these were only random airborne flavors in fleeting combinations that sparked her memory. She recognized other, more persistent cues as signs that guided her across the sea.

She used her sense of smell and, indeed, all her senses to hone the pursuit of her particular niche. She and her kind had mastered the unseen pathways across the water so well that they had divided the ocean's resources into more than sixty niches—one for each species of petrel. Many of those species count their numbers in the millions, even though they are unknown to land-loving eyes.

The land increases in its living diversity as the latitude decreases—the plants and animals flourish toward the equator and diminish toward the poles. That gradient is reversed below the surface of the open ocean. The higher-latitude seas, where the colder water dissolves more oxygen, teem with life, while the tropical sea-lanes are comparatively barren.

Above the waterline the high-latitude islands are pitiless deserts, but just beyond the shore lies a bountiful sea. These islands are perfect places for seabirds to nest—remote, hostile to terrestrial predators that might try to wait out the freezing, barren winter, yet well situated for offshore foraging when the time comes to rear chicks.

As she rolled with the low swells, the scent of the Antarctic Convergence came to the petrel—one of the many boundaries that demarcate the open water. The convergence marks the line where cold currents spread out from the Southern Ocean to meet the warmer waters to the north. The cold current carries debris and plankton, and where it subducts beneath the warmer water, it leaves the rafts of this buoyant flotsam to accumulate. This line of countless tiny buoys releases a particular bouquet of aromas across the temperature gradient. Its margins are outlined in populations of krill, a feature on the sea well known to the petrels.

The petrel's focus returned to closer waters when she caught the smell of ink. Somewhere nearby a battle was going on involving squid. She dipped up a billful of water, and the taste of ink—stronger than the odor—confirmed her suspicion.

The breeze shifted and the tangy whiff of a distant nesting island floated past her nose. Overpoweringly strong at its source, that scent was detectable over great distances of dilution through the maritime air currents. It was a suggestive fragrance for a petrel. She would have to investigate the notion of nesting one day.

A wormlike coil, pallid in the starlight, floated up beside her. She dipped it up and tasted it: arm of squid—one of her preferred morsels. She was so at home on the ocean now that her meals were finding their way to her.

The scent of the bird island lingered on the water. She had mastered the skills required to support herself, and soon she would have to move to yet a higher level and learn to fish even more efficiently—to feed herself and a chick twice as hungry as she. She would find a mate—perhaps the one-footed male would suddenly look different to her when he was standing on the rocks of a nesting island. Petrels mate for life, and pairs raise new chicks every year.

But young petrels often do not even visit nesting islands until they are three years old. Now, her attention was still on the nuances of her immediate surroundings. She paddled slowly and watched for telltales she once might have missed. In time she would understand what the slightest sound or flavor had to tell about her water world. She would learn to persevere where a yearling of her kind would have starved in tepid waters bereft of fish in an El Niño summer, or perished in the dark of an austral winter storm. Eventually, she would come to see the monotonous plain of the open ocean as a landscape—complex, shifting, offering a new choice of paths to prosperity each day.

Science Notes

One of the differences between the birds of the land and the birds of the sea is in their senses of smell. Land birds (with the exception of the vultures) appear to depend heavily on visual cues and are thought to have a diminished sense of smell. In contrast, birds of the open ocean—of which there are just as many or more than on the land—depend on their olfactory senses. One dominant family of open-ocean birds is the Procellariiformes. They include the petrels (65 species), the shearwaters (20 species), the albatrosses (20 species), and the fulmars (4 species) (Brooke, 2004). They can all be called "tube noses" because of the long covered nostrils on their beaks. Their enhanced sense of smell allows them to detect the subtle olfactory cues that guide them across the waves (Safina, 2003; Nevitt and Bonadonna, 2005). For example, drifting algae (phytoplankton), when fed upon by animal members of the plankton (zooplankton), release certain volatile chemicals. The scent of those chemicals (e.g., sulfides) guides the next-higher order of predators (nekton, animals that do not drift with the currents but can cut across them) over the sea to where the feeding activity is (Nevitt and Haberman, 2003).

The smaller members of this group, the petrels and shearwaters, characteristically hold their wings above as if in a hover, extend their feet down, and run across the waves when foraging. Many species of petrel and other seabirds have learned that their foraging is improved if they follow dolphins or tuna (Au and Pitman, 1986), false killer whales, or fishing boats. Convergence zones accumulate buoyant material—living plankton or floating debris that collects at down-welling boundaries where currents collide head-on and dive beside one another (Dandonneau et al., 2003); these zones of flotsam are highway lines for those mobile, open-ocean (pelagic) creatures that live far beyond the shore and far above the (benthic) creatures on the bottom.

Seabirds predominate in "the water hemisphere" of the globe (Editors of *Life* and Rand McNally, 1961:425), which is centered on a pole that rests at 47 degrees south and 173 degrees east (just east of New Zealand). This hemisphere is more than 95 percent covered by ocean; it includes the vast South Pacific, much of the Indian Ocean, and the Southern Ocean that surrounds Antarctica.

An island leaves a wake if it stands in a five-knot current, just as does a whale moving at five knots—though the island's wake is of greater magnitude. The eddies in the margins of an island's wake stir up the lower layers of the ocean, levitating nutrients to the surface. The surface waters of the midocean are nutrient poor—the trace essential elements there have all been sequestered by the plankton. For example, iron salts are missing from the surface layers, and iron is limiting for planktonic growth (Coale et al., 1996). Soluble forms of iron from the lower levels of the ocean, when carried to the surface by mixing disturbances, unleash a bloom of phytoplankton (De Baar et al., 1995).

Air breathers like the birds have the potential for much greater levels of aerobic metabolism than water breathers. The concentration of oxygen in air is thirtyfold greater than that dissolved in water; air is easier to inhale than water is to pump across gills; oxygen diffuses a million times faster through air than through water (Steen, 1971). The seabirds brought this advantage—their development of lungs—to the competition when they returned to challenge the fish for dominance of the surface layers of the ocean.

The distance to the oceanic horizon as a function of the viewer's altitude is calculated as the square root of H (height above the water, in feet) x 1.16 = D (in nautical miles). The area surveyed from this vantage equals $(\pi)D^2$.

References

Au, D. K. W., and R. L. Pitman. 1986. Seabird interactions with dolphins and tuna in the eastern tropical Pacific. *Condor* 88:304–17.

Brooke, M. 2004. *Albatrosses and petrels of the world*. Oxford: Oxford University Press.

Coale, K. H., et al. 1996. A massive phytoplankton bloom induced by an ecosystem-scale iron fertilization experiment in the equatorial Pacific Ocean. *Nature* 383:495–501.

Dandonneau, Y., et al. 2003. Oceanic Rossby waves acting as a "hay rake" for ecosystem floating by-products. *Science* 302:1548–50.

De Baar, H. J. W., et al. 1995. Importance of iron for plankton blooms and carbon dioxide drawdown in the Southern Ocean. *Nature* 373:412–15.

Editors of *Life* and Rand McNally. 1961. *Life pictorial atlas of the world*. New York: Time; Chicago: Rand McNally.

Nevitt, G. A., and F. Bonadonna. 2005. Seeing the world through the nose of a bird: New developments in the sensory ecology of Procellariiform seabirds. *Marine Ecology Progress Series*, 287.

Nevitt, G. A., and K. Haberman. 2003. Behavioral attraction of Leach's storm petrels (*Oceanodroma leucorhoa*) to dimethyl sulfide. *Journal of Experimental Biology* 206:1497–1501.

Safina, C. 2003. *Eye of the albatross*. New York: Henry Holt.

Steen, J. B. 1971. *Comparative physiology of respiratory mechanisms*. New York: Academic Press.

Sturgeon

The sturgeon had lived a great, long life. She was one hundred years old, yet she showed no signs of her age, other than her size—she was twenty-five feet long, and out of the water would weigh a ton. She had lived long enough to see the effects of geologic processes that move so gradually as to go unnoticed by creatures of lesser longevity. She was now witnessing such a process—the change in the course of a grand old river.

One sign of the change was in the flavor of the river water—it carried a foul aftertaste, an anaerobic overtone of stagnant silts and decomposing reeds. Year after year, this river's flow had been steadily decreasing—one day it would cease entirely. The motion of earthquake faults in the rocks far inland were diverting its path—its delta would eventually disappear into the wide crescent of Monterey Bay.

The sturgeon had been migrating up this river every few years for most of her life. But now she lay motionless on the sea bottom twenty-five feet down, stymied by indecision. Above her, the fan of warm freshwater spread out across the ocean surface. She had already crossed the barrier between that layer and the salt water twice today and had turned back each time. She turned to cross it once more and was again seized by the urge to retreat. Changes in the inland watercourse could threaten the success of her spawn.

She lived for this migration back up the river of her birth. That river extended for hundreds of miles, cutting through the coast mountain ranges. She had followed it to the mountains dozens of times, but each successive trek across the broad delta had been more difficult than the last. Over the years she had grown in size and strength, but the river had done the opposite. Its ancient canyon was the drainage from a great central valley, but its channel had become shallow and clogged, the sluggish current no longer cold.

She cruised along the beds of kelp parallel with the shore, mapping the gradient of riverine flavors that floated across the submarine slope and disappeared seaward. Instinctively she turned toward the shore at the peak intensity of freshwater cues but soon broke off the approach. She repeated

The secret is out. A flounder may have been discovered, even though it is buried.

the search pattern again and again but broke away sooner each time. Eventually she turned sharply and fled west toward the open ocean.

The sturgeon descended along the walls of a yawning chasm, comparable in grandeur to the Grand Canyon of the Colorado, yet unknown above the water line. Her river had excavated that gorge during an epoch when the sea level had been lower. At that time—before its strength had begun to ebb—the river cascaded four hundred feet to the beach, excavating a spillway deep into the cliffs. She glided silently above those now drowned bluffs, along a deepening blue defile that continued a mile farther down.

She was oblivious to the midwater distractions—the school of opalescent squid retreating backward ahead of her, the seals trailing streaks of bubbles above. The pulsing skirts of the Pacific jellyfish did bend her course—they would reach out invisibly and sting if passed by too closely. She gave them a wide berth while keeping her senses focused where they were most accustomed—on the bottom.

A pallid, semicircular coil lay in the ooze ahead. She slowed to a cautious glide, nearly stopping as she approached the object. Her great weight was exactly balanced by the buoyancy of her swim bladder, so in the water she was weightless—easily maneuvering herself to within inches of her target. She came to a hover above the motionless form and investigated it with sensitive barbels that dangled from her snout. Dead eel. Her mouth expanded down over the creature and she sucked it up, along with those scavenger crabs that were reluctant to release their grasp.

Before she moved on, she noticed something else nearby. To most creatures, there would be nothing visible in that direction amid the sparse debris on the seabed, but the sturgeon could see through the mud. Using a field of conductive pores on her snout, she could sense electricity. As she approached, the array of sensors on her head drew a picture for her of an enlarging oval on the bottom, outlined in electrical currents coursing through its own nervous system. Though this patch of sea bottom appeared no different from the rest of the featureless silt, she was guided by a visualization that appeared to show a buried right-eyed flounder.

She glided to a perfectly balanced stop above the chosen spot. Her barbels quivered just above the mud as her protrusible mouth extended. Then quicker than the eye could see, faster than a startled creature could bolt, she sucked up the buried flatfish, together with an equal weight of sand.

The sturgeon pursued a course up the coast, directly against the current. She left her wide crescent bay and cruised north along a coastal shelf less and less familiar. Though not consciously following a scent trail, she was nonetheless driven by the upcoming spawning season. Textures of the bottom blurred into the continuum of motion as she bent her strength against the perpetual resistance.

She was in need of a river, so she pressed ahead through the oncoming current, probing for riverine scents. Diluted hints of the olfactory sensation she sought flickered across her senses, only to vanish into the background of ocean flavors again and again.

After a hundred miles and several days she came to realize that a gradient of riverine flavors was actually emerging above the subliminal level. The cues she sought were here, and growing in strength to clearly perceptible concentrations. She tasted the acidity of pine needles decomposing in quartzite soil, the ions leached from granite cliffs—it was the bouquet of the mountains, diluted across the familiar background flavor of the coastal ocean.

The scent trail diverged from the main coastal current. It led her across a shallow gulf that moved the coastline well eastward, away from the contours of the offshore shelf. Soon the compelling flavor of freshwater grew unmistakable, issuing through a narrow passage in a steep range of mountains.

But the flow was not from a river delta—it was much too strong. Though the rush of scents enticed her onward, the closer she approached the opening in the shoreline, the stronger the opposition became. A momentary fatigue settled upon her and she glided to a stop on the bottom, stymied once again. The irresistible current pressed her against braided sand, bathing her in its bouquet of flavors.

Twenty million years is but a small fraction of the time the sturgeon and her order of ancient fishes had plied this coastal littoral. But during that time, a great earthquake fault had reshaped the seashore. Along the northward arc of the fault line, between the wide crescent bay she left behind and the young estuary she now discovered, the fault had branched. A deep spur split from the main discontinuity, and the two fault lines advanced northward in parallel. Between these twin rifts, a basin had formed.

One range of mountains east of the spur line and a second range west of the primary fracture pulled apart as the margins of the fault blocks shuddered past each other. The basin between the two fractures subsided, sinking toward sea level, gradually forming the setting for a great inland bay.

The sturgeon awoke from her pause with the sense that the current pushing her onto the bottom had slackened. She realized that it must be the tide, and it was now reversing. Though she had lived with the tides all her life, she had never had them focused against her like this. This must be the mouth of a great estuary, an onshore arm of the sea. Nonetheless, the flavors of dry land carried on the tidal flow assured her that a river ran through it.

She knew that the closer she swam to the bottom, the weaker would be the opposition of a head-on current. And she was adept at swimming close to the bottom—she was shaped with a curved profile above, and flat underneath. She rose from the sand and put her strength against the resistance, cruising a foot above ground, and pressed ahead across an entrance through the mountain wall.

Following the floor of a tidal channel for four miles, she prevailed against the head current. The cliffs that rose on either side shown sunset red through the water, but the north wall was slashed with a tall, dark, sea green scar where a freshly ablated obelisk of stone had slid away, revealing the unweathered interior of marine basalt. Shattered fragments lay all along the waterline where the channel undercut the sheer face straight above.

The canyon walls eventually spread away and the sturgeon found herself in an estuary too broad for her to know the measure. The water was filled with other fish—bass, perch, bluegill, sturgeon. The smaller fin were plundered by otters and ducks diving from a surface littered with paddling rafts of their kin.

The sturgeon's senses swam with the assault of flavors. The prevailing currents were masked by the tidal flux—salt water mixed with fresh with no apparent direction to the gradient. Mazes of channels curved around islands.

Dryland scents were strong in flows that carried stream runoff; upland plant matter drifted everywhere. Shoals of shellfish mantled every outcrop. Plankton clouded the waters, anchoring the bottom of the food chain amid an abundance of nutrients that are scarce in the open ocean. The sturgeon could taste these elements—iron, zinc, copper. But most distracting was the taste of the distant mountains, more prominent in this mélange of flavors than it was offshore. She set off to find its direction.

Gliding from the tidal channel, she cruised across the algae green silt. Immediately she noticed something off to the side and bent her body nearly in half as she changed course to investigate. There was nothing to see on the bottom, yet her electrical sense guided her toward an outline ahead, perhaps something buried in the ooze. An angular profile enlarged in her electric sight, resolving into a square shape as she smoothly glided to a stop above it.

Suddenly the electrical image flared in brightness and the square expanded, spreading rapidly to cover the entire surface below. She felt a tingling sensation across her snout and involuntarily rose away from the bottom, tipping slightly in disorientation. She found herself angled sideways to the wind-driven surface current, so she regained her balance and turned into the flow. As she increased her headway, her dazzled electric sight slowly began to clear.

The skate buried in the silt had not moved as the ominous shadow had descended upon him. He had done nothing more than reflexively pulse a few volts per second of nervous electric current into the salty water. The charge had blinded the sturgeon's ability to see the skate's nervous system through the mud—saving his life.

For generations of sturgeon, the great spawning river had flowed along the inland margin of the mountains east of the twin faults. But the land was rising along the river's southward path, impounding its flow. At the same time, part of the mountain range was sinking into the broad basin forming between the two faults to the west.

Every year the river appeared to rise higher along the inland edge of the descending section of the range. Eventually, during a high spring flood that turned the great river valley into one of the world's largest lakes, the waters spilled over the lowest saddle in the ridge profile—and began to cut a canyon through it as they cascaded into the subsidence basin to the west.

The basin was filled by a lush redwood Eden—fog-filled in the morning, warm and humid by midday—surrounded by hills forested in hardwood brush and live oak. Its deep center was home to elk and turkey, ground sloth and saber-toothed cat. A watercourse with shores choked in ferns and vines meandered through its center, eventually flowing out through a notch in the western mountains, over a bridle-veil falls above a beach on the ocean.

Over the years, the side shoot of the great river to the east carved an ever-deeper notch through the saddle in the subsiding mountains. A furious, muddy spillway poured through in the spring, swelling the flow in the basin's drainage threefold, straightening the stream's meanders, separating the banks, and deepening the notch through the mountains. As the flow from the east grew strong and permanent, the pulsating flood wore down the western wall to the ocean at the peak of the spring runoff.

But then the storms began to diminish. An ice age was ending, and the sea level rose. The tide line crept up the flank of the coastal range and swamped the cascade at the river's mouth. The sea reversed the flow through the western notch, and the redwood forest that once gave its deep shade to the glens and hollows of the broad basin was drowned in salt water. The pulse through the notch was now driven in both directions, twice daily—by the ocean's tides.

It took the sturgeon days to find her way out of the estuary. She explored channels lined with fallen trees ten times her length—their roots rotted in seawater and their trunks rolled down across shorelines rising around islands that were once hilltops. These trees had grown for centuries and would lie preserved in the silt for centuries more. The steeper slopes of the submerged terrain were eroding underwater, the ground shifting to fill in flooded lake basins and drowned gullies, gradually leveling the bottom of the estuary to flatness.

The sturgeon explored a long southerly extension of the bay that dead-ended in tributaries heavy with the taste of chamise and ashes, dust, and dry, fire-cycle lowlands. No great snowmelt river entered there.

Eventually she found the channel of a drowned riverbed snaking from the north through waters opaque with suspended silt. She felt her way through this submersed canyon as it curved east, and there the river revealed itself to her through the scent of its strengthening flow. Mountain walls again drew in upon her as the estuary narrowed and the current increased.

A full moon rose above the springtide as the sturgeon entered Saddle Canyon for the first time. Darkness was falling in the already murky floodwaters, but she was unimpeded, navigating by her sense of distant touch. She carried a pattern of pressure sensors in rows extending down her flanks. These allowed her to feel the pressure waves in the water, much as creatures of the air hear sound waves. She knew the speed of the water flow by the pressure of its turbulence, knew of the movements around her by the pressure waves radiating ahead of her motion.

Much of what she felt was a reflection, the return of the pressure generated by her own headway. This was a form of echolocation. She had learned to listen to her environment as it reflected pressure back to her—strongly from nearby surfaces, weaker from more open water. The outlines of the dark landscape were drawn for her in echoes of pressure waves bouncing all around.

She sensed the entire channel, from the hard clean boulders lining the bottom all the way to the strand line above. These images, together with her sense of the strength of the current—strongest where the flow had established the channel center—showed her the path to follow, even though only darkness fell upon her eyes.

The current stiffened as she passed beneath Saddle Canyon's young cliffs, where the whole watercourse narrowed now to less than fifty feet. Crossing the rocky bottom, the sturgeon encountered the biggest boulder in the river rising squarely above her in the channel center. Though it reached nearly to the surface, she knew by the pressure of the current flowing above it, and the lack of flow on either side, that her path was directly across this rock. She had the strength and the determination to move through the constriction despite the power of the flow focused there against her.

The rising moonlight was beginning to take full effect on the rapids when the sturgeon's back broke the surface. Her form appeared as a glistening arch standing in place on the water above the submerged obstacle ten yards from shore. The arch of her back rose higher as the wider parts of her midsection flowed over the rock. Her dorsal line of spikes glided along the stationary arc, slicing up out of the water, smoothly parading along her spine and hissing back below the surface against the current upstream, like a scaly water wheel rising in midstream.

A pack of wolves on the far shore stood back on their heels, hackles up and bristling, eyes wide and snarling, as if they expected a plesiosaur to rear up in midwater. On the near side a pair of brown bear cubs hightailed in the opposite direction, disappearing into the forest. Their mother stood her ground, alternately raising her snout to sniff for the scent of danger and glowering across the water at the unknown menace.

The sturgeon spent the night negotiating the current among the pillars fallen from the rim of Saddle Canyon. When the turbid waters brightened with daylight, the resistance eased and she passed into a much wider realm. There, a sense of recognition dawned upon her—this was again her home river.

Its familiar bed teamed with the blackfish, hitch, and suckerfish she had always known. She did not stop to wonder why she was already halfway up it. Instead, she pressed her snout into the mud and dug up the bottom. Using her barbels to feel through the murk of suspended sediments, she found a few worms to eat, and some shellfish now lying exposed on the bottom. She continued to excavate, consuming most of a bed of scallops along with a number of clam-shaped rocks.

As it was most springs, the broad central valley was transformed into an inland sea. The sturgeon was able to stray far beyond the primary water-way, cruising above what was and would soon again be bunch-grass prairie. Male sturgeon often followed in her wake.

She bent her path out across the shallow flooded plain to investigate a swollen river cut through the golden sandstone rock of the mountains rising west of the valley. But as the land slowly rose and rapids began to thrum in the water, the sturgeon lost headway and settled lower among the drowned snags of full-sized oak trees.

The river washed its flavors over her—suspended traces of what once was lingered in the water—and a flood of recollections returned. This was the lost tributary of her birth, flowing again after being closed for most of her life. She was now following the same river she had retraced at age eleven on her first naive spawning run. She had been a small fry then, only eight feet long. Now, as she lay motionless at the bottom of the channel, the innate memory that guides sturgeon back again and again to the same river was awakening, reminding her of how she knew this path.

The water carried the dimly familiar flavor of a lake, and memories of the futile years she had spent in it. Long ago, the last time she had ascended a river to that lake, she had found her return to the sea blocked by a wall of fresh volcanic basalt—leaving her landlocked. During those freshwater years she had lost weight—there was not enough life in that lake to support a fish of her size. She had captured some of the slower fish that had entered the lake, but others of those slow fish grew over the years and eventually became recognizable as sturgeon—she had been eating her own fry that had followed her downstream after hatching and were trapped in the same impound.

She had eventually escaped the land lock—struggling down a narrow outflow that led her on a shallow trek through the forest, over semi-submerged boulders, leading to a waterfall that nearly killed her, and finally back to the sea.

The sturgeon had now slowed to a stop. She rested on the bottom, below a thicket of drowned snags in the narrow channel of the sandstone canyon. Finally her great mass rose straight up, filling the water with a blizzard of drifting black twigs snapped from the branches on either side. Though the channel was not wide enough for her to turn around, she continued in motion, jack-knifing upward across the narrow confines to propel herself toward the shore.

The herd of horses resting beside the river rose to its feet in unison as the waters parted and the sturgeon beached herself directly before them. Chewing of cud ceased in midstroke. A dozen pairs of eyes gazed down at the vast domed head slanting up through the surface. Above its wide-set eyes, five parallel lines of armored spikes glinted along the spine and flanks of a body more massive than all of them combined.

She was much heavier half awash, resting unbuoyed on the sand. But the sturgeon nonetheless found the power to explode out of the water, exposing her full length amid expanding spiral beads of spray. She ponderously jack-knifed in midair, landing against the wet sand with a clap that echoed along the canyon walls. The horses scattered—running in every direction as fast as they could—while the sturgeon, now facing downstream, disappeared back toward the main river channel.

The temporary inland sea knew no tides, no swells. Its mirrored surface covered the great valley, unbroken between the eastern and western mountains but for a double file of oak treetops snaking south, then west. This hedgerow was the only sign of the river corridor. The riverway itself had vanished weeks earlier under a continuous breach across its own banks.

The flood spread through the foothills east and west, where it met its swollen tributaries in ravines that cut farther back into the mountains. Runoff from every province around the valley found its way through these canyons onto the overflowing floodplain below.

The sturgeon's home waters in the Pacific were no farther offshore from the coastal mountain ranges than was the channel of this central riverway from the water's edges against the surrounding foothills. In her ocean, however, mountains never completely encircled her—infinite ocean always stretched away to the west. Here, open water disappeared over its own horizon only across one thin narrows to the south.

The sturgeon swam through a perpetual turbid twilight against the current at the bottom of the central channel—occasionally deviating to explore beyond the rows of great tree trunks that bounded the river corridor. There was less head current out above the shallower inundated prairie. But she could lose compass there and stray from the main channel, so she always gravitated back to the deeper straits.

There was also less forage beyond the channel basin. She sometimes encountered schools of black catfish fry contracted into floating balls at her approach. Their strategy provided defense against heron, salmon, and other predators that needed to focus on a single target before striking. But their schooling response merely aided the sturgeon in sucking them up in greater numbers.

She had put nearly four hundred miles of valley floor behind her in the weeks since she left the brackish estuary—and was ready to go on for a thousand more. But the waters were chilling, the riverbed rising, and the current increasing. The splayed footings of cottonwood were replacing the clean dark oak trunks on submerged levies. Muddy sediment on the bottom was gradually giving way to smooth rounded stones, their rusty colors bright under rippled highlights penetrating the water.

The snow was getting closer. Tastes of the acrid ionic forms of iron, lead, potassium, and aluminum—the primary weathering salts of the mountains—were increasingly prominent. These ions were never tasted in the open ocean, where they were immediately sequestered by the plankton. Conversely, the sodium flavoring of seawater was no more than a receding memory here.

Granite facings closed in on the channel as the river rose out of the foothills. Steep walls began to curve sharply, narrowing almost to the sturgeon's own width in places. Soon, she would be able to advance no farther across the rising terrain.

At the head of a deep basin in the riverbed the sturgeon came upon a marble wall shrouded in sun-bright columns of bubbles. The current vanished into random fluctuations as she approached the blockade. She slowed her powerful stroke and drifted backward, watching the spinning eddies merge into larger vortices, eventually reforming the main current she had passed through moments before. Any further advance was blocked.

Here the rapids began and her journey upstream ended. The sturgeon was a creature of the onrushing flow, but the water fell vertically from an overhang into this hole, the first of many such obstacles. She was not going to crawl over exposed boulders; she had no need to do so. Settling on the gravel bottom, she spent the rest of the day beneath a procession of steelhead that ascended the stream in twos and threes. Each one accelerated with a flick of its tail and disappeared through the surface above her at the foot of the rapids.

The sturgeon familiarized herself with every inch of this last deep-water segment of the river below the falls. This was bright water, revealing its granite surrounds in shining clarity. It was the opposite of the translucent screens of suspended silt she lived behind most of her life. Here her sensitive barbels could detect no trace of the sodium chloride that usually flowed at equal concentrations through her gills and through her veins.

After several days of exploration, she abruptly lost interest in her investigations and settled to the bottom to wait out the night. With first light, she rose through the water column and moved upstream to find a region where the river's great strength was focused. There, she came to the surface to face a constriction in the channel through which the pressure of the current was amplified by a steep, descending gradient. She matched the force of the current with continuous powerful strokes that rippled down her entire length—holding her position beside the polished granite walls.

Suspended in the swirling, buffeting torrent, she relaxed, strengthening and lengthening her stroke. Finally, she began the work of dropping her dense masses of black eggs. They emerged in numbers that transiently cast the bottom in shadow. A million. Two million. More.

She worked smoothly, without interruption, contributing her part to the reversal of the order of the flow of life from the land, through the rivers, ultimately to the sea. She was returning to its primordial upland sources that which she had collected from the salt water—refashioned in the form of her spawn.

The eggs raced into the swift waters, swirling through the twists and helices of current that buoyed their descent over the streambed. Many were still suspended as they were carried out of sight down the channel. Between their dispersal from one another and their first encounter with the rocks on the bottom, they metamorphosed: hard and smooth on release, they became soft and sticky in seconds. They executed their crucial transition in texture with unerring timing. Thus they minimized clumping, yet moments later they clung securely to anything they touched in the interstices of the riverbed.

The current pursued every avenue through the gravel riffles, no matter how small. It divided itself through tunnels between the stones, across the undersurfaces of ledges, and into crevasses in sunken snags. The eggs it bore found protection in all these places. There they spent a few minutes more, modifying their surfaces further to accommodate fusion with waterborne sperm. And then they waited.

The sturgeon had noticed no males in the area, but as soon as she finished shedding, two slender shapes appeared slightly above her in the water, one on either side. They were much smaller and sleeker than she, only ten and eight feet long—a third of her length—yet they kept station effortlessly in the driving current. The three hovered in the slipstream, passing two, three, four minutes.

Then the cloudy issue of milt appeared synchronously behind both males. They shuddered and flinched as they shed, sending off countless flagellated cells that would watch their entire free-living life pass in a minute or two. Most would be borne away in midstream, out of reach of their goal.

The shadows slanting down through the water from the pines that reached out over the surface came into sharp relief as the turbid cloud expanded, carried with the current, filling the entire channel. Above it all, three sleek silhouettes hung motionless, suspended in formation, floating upstream.

Her spawning had coincided with the highest surge of spring's flood crest, and now the receding waters carried the sturgeon south out of the mountains, back to the valley floor. The meanders she had followed upstream had been straightened by the flood. Over the years the river was continuously snaking back and forth across the plain, silting in its bed, overflowing into new channels—filling a steep valley with layers of sediment as flat as the water surface above it.

As she passed, dry land emerged behind her. The retreating waterline stretched perpendicular to the river, across the valley floor. Animals descended from the foothills to explore the newly rinsed and fertilized prairie she had previously explored from the other side of the receding shore.

An army of wading birds—cranes, egrets, curlews, sandpipers—stalked the moving margin that paced her trek toward the sea. Undulating rows of geese divided the sky above into herringbone patterns that progressed in unison toward the northern horizon.

Aquatic predators and scavengers descended on the upper reaches of the river in search of sturgeon eggs. But even if they found most of them, the number of sturgeon larvae that remained hidden in the rocks would produce more than enough new fish to explore all the new habitat the floodwaters could create.

As she slid past the cutoff toward Saddle Canyon, the sturgeon kept to the main river channel, as had always been her habit. But a few miles farther she found the current slackening, noticeably weaker than it had been at this point in her last spawning migration five years ago. The central channel had silted in even more. The water grew warm as the river flattened out and slowed.

A rising mountain range was closing this arm of the river forever. A broad black monolith that would one day be named Mount Diablo was enlarging squarely across the waterway—its massive solitary ridge shoving the flow aside. Aided by the upstream diversion through Saddle Canyon into the deepening basin to the west, the rising landform here was killing this southern arm of the waterway. Marshes encroached from both shores toward the languid central channel. The sturgeon settled to a stop on the bottom. Then she rose and spun around to press upstream against the current once more, retracing her path back to the Saddle Canyon cutoff.

She found the canyon channel rearranged since her passage just weeks before. The torrent had further undercut the walls, widening the gorge, tumbling new barriers from above. The river had improvised new courses across its changing bed.

The young estuary below the canyon seemed to team with even more activity than before. The sturgeon settled motionless in the bottom of the deepest channel and waited while the leviathan shadows of a pair of humpback whales darkened the waters above. And then a school of millions of languid herring stretched off out of sight ahead. Its members aligned along unseen cues and condensed into braided columns. Tenuous arches materialized across the width of the extended community of fish; then the flowing networks dissolved as the individuals wandered from the pattern.

The entire leading edge of the school now bowed into a growing concavity. The sturgeon had caught the scent of the open ocean. The main estuary currents merged here, forming an avenue outlined in flecks of pyrite and mica borne from the mountains, glowing golden in the slanting evening sunlight. The broad school of herring parted into a continuous, slowly rotating tunnel, yielding to the irresistible force that impelled the sturgeon to the west—riding the outgoing tide past the red-rock gates towering above the outlet to the sea.

Though she was a hundred years old, she was as strong and vigorous as she had ever been. She would live for another century and adapt to the geological evolution of a landscape around her.

She was a creature of habit and had memorized the channels that contained the rhythms of her migrations. The longer she grew, the more sensitive grew the rows of pressure sensors running the full length of her flanks. They showed her the changes recorded in the passing landscape from one migration to the next. They felt the shockwaves of even the slightest earthquakes flying just below the surface, and they recoiled at the furious, concussive passage of the nascent tidal waves spawned by major movements on the coastal faults.

She recalled what had come to pass each time she encountered reminiscent flavors—barely perceptible reminders that became dilute and faded if pursued—suspended traces of what once was, lingering in the water.

Science Notes

Great white sturgeon, like the one described here, linger in the water now only in our imaginations. That species rarely reaches even six feet in length in modern times—a consequence of blockades in its spawning rivers and of overfishing, including the harvest of females for their roe (caviar). We will not encounter such a magnificent fish anytime soon migrating upstream, nearly filling the streambed with its size, pressing through the forest against the height of the springtime flow. White sturgeon (*Acipenser transmontanus*; Moyle, 1976) have lived in rivers and along the coasts of the Pacific Ocean for nearly two hundred million years, since long before the time of the current tectonic epoch. These fish rely on their lateral line pressure sensors, as well as their sensors of faint electric current (Coombs et al., 2002), for information about their surroundings.

This tale is set against the epoch of the movement of the San Andreas Fault in western North America. Tectonic change has resulted in multiple closings of the mouth of the great river that drains the vast inland Sacramento/San Joaquin Valley—once a prime sturgeon-spawning river. The river has moved its delta north with the fault, from positions in what is now Monterey, to Colma south of San Francisco, to its current position at the gateway to San Francisco Bay. The story also alludes to the lava-flow closure of the outflow of Clear Lake, north of the bay; this episode would have landlocked the migratory fish there at the time.

Sturgeon are faithful to their spawning rivers. They live long enough under natural conditions that, should the mouth of their favored river close, they have to adapt—as does the fish in this tale. Saddle Canyon here represents the Carquinez Strait east of the San Francisco Bay. That river path has been cut by the Sacramento River in recent geologic time as the river has been diverted by the changing terrain. The river's southern path to the sea was closed by a rising landscape, while the land to the river's west slumped. That slump, between the

San Andreas Fault and the Hayward Fault, produced the depression now filled by the estuary of the San Francisco Bay (Sloan and Karachewski, 2006), the latest outlet for the waters from the Sacramento/San Joaquin Valley drainage. When sea level falls from its current maximum, that bay will drain, exposing bottomland suitable for the return of a deep redwood forest.

References

Coombs, S., et al. 2002. Information processing demands in electrosensory and mechanosensory lateral line systems. *Journal of Physiology* 96:341–54.

Moyle, P. B. 1976. *Inland fishes of California*. Berkeley: University of California Press.

Sloan, D., and J. Karachewski. 2006. *Geology of the San Francisco Bay region*. Berkeley: University of California Press.

Life in the Sky

The sea is tinted green with plankton, streaked vermilion with floating krill, crossed beneath the surface with silver flashes of herring, and shot through with the plunging forms of blue-footed boobies—a feeding frenzy is churning the surface. Above it all, currents of warm air rise off the tropical ocean.

The commotion summons an apparition—a black silhouette from the haze on the far horizon. It is the quintessential embodiment of this water world—the magnificent frigatebird. Its effortless transition across the seascape is proof to sailors that they have reached the tropics. Distant shores move past it, but this scissor-tailed kite approaches motionless—a shadow changing only in perspective while the open ocean slides away behind.

While the winds aloft buffet everything else, the frigatebirds sail serene, spinning the same spirals through the sky that the most accomplished high-flying eagles and condors follow above the coastal mountains. But the signature bent-wing figure of the frigate is not found over mountains, or far inland, or anywhere far from the low latitudes—the frigatebirds fly only over tropical oceans, where the waves are warm enough to generate the thermal updrafts on which they soar.

Only the albatrosses have longer sea-gliding wings, but their flight paths tend toward the stormy latitudes. They do not soar on thermals—thermals do not rise from their colder oceans. They slice across tempestuous seas only a few feet above the water, cutting through troughs, slipping over the whitecaps, slope-soaring along the faces of the waves while nothing soars farther above them.

The instantly recognizable figure of the frigatebird is unique to the tropics, sculpted by the convergence of circumstances there. These gliders lock their wings in position and hold them steady for hours; they use little energy and generate little muscular heat. The tropical warmth helps maintain their body temperature when their lightweight, high surface-to-volume form loses heat to an afternoon thundershower. But the warm tropical water works against them—precluding the production of the abundance of fish found at the albatross latitudes, where oxygen dissolves to greater levels. The frigates' gliding way of life is an adaptation to this shortage of food—it optimizes the bird's capacity to wander far and wide. Frigates climb thousands of feet to scan the curved oceanic horizon from the vantage of their lazy circles—looking for concentrations of prey that rise into view from tropical seas only rarely.

These birds have taken their specialization in soaring to its aerodynamic extremes. They have the lowest wing-loading ratio of any gliding bird—a surfeit of surface area in proportion to their weight. Their feathers are dry and light—not oiled enough to repel seawater and allow them to float—so they never dive into the waves or even alight upon the surface.

The exaggerated aspect ratio of their wings—narrow yet more than seven feet long—maximizes the speed they can maintain when gliding at their minimum sink rate. Their sharp wingtips direct the drag generated by their motion far out and way from the main portion of the wing. They slide through the sky straight and flat, yet they bank easily into updrafts. They are one of the few birds that can survive a hurricane by riding it out on the wing.

Frigates may follow schools of porpoise or tuna, watching for smaller fish to be driven to the surface. These birds cannot enter the feeding frenzy directly—they live along the margins. A needlefish that shoots from the water might confound its pursuers below, but where it skips along the surface, sculling with its tail, it exposes itself to the frigatebird. The black raptor drops from above on half-closed wings to pick off any airborne morsel that enters its dominion.

With the grace of a kestrel, the frigate times the trajectory of the flying squid, or the glide path of the flying fish, and snatches away its prey just before it reenters the water. Only its hooked beak gets wet when the bird skims the waves to pluck a catch from close beneath the waterline. Thus, the herring that feed on floating krill will dash up for an attack, strike, and jackknife to dash back down all in the same motion, minimizing their exposure to the skimming fisher.

The boobies are the aerodynamic opposites of the frigatebirds. Their compact bodies are stout enough to endure the pounding of repeated headlong crashes into the waves. They perform this maneuver all day long, riding the momentum of their plunge twenty feet below the surface, chasing the silver quarry they target from above. Then they reappear on the surface, toss their catch back into their crop, and run along the water, pumping their wings to regain the air. The inertia of their heavy flight holds their

Crossing the Vermillion Sea. A magnificent frigatebird
navigates toward a roost beneath the mountains of Baja California

path straight, their turning radius wide, and their ascent profile long and
shallow.

The sight of a booby with a bulging crop is an invitation to the frigatebird. This mute predator was named Man-o'-War bird by the corsairs of
old because it too is a pirate. One way it sustains a life that never touches
down is by augmenting its diet with an occasional in-flight meal of warm
fish.

Each booby is constantly aware of the frigatebirds drifting far above.
But on occasion it loses sight of the high-altitude patrol, only to feel itself
engulfed by a wave of lighter air—accelerated by the pressure of one of the
imposing black intruders following right behind it.

The frigates always have the aerial advantage over their victims. The
booby looks back to find a brigand with a wingspan nearly twice its own,
riding behind—uncomfortably close and matching its moves exactly. When
it turns its head to cry out at the attacker, it kills its own forward speed, allowing the pursuer to close even nearer—the frigatebird wants the booby's
catch.

For the booby, evasive flight is futile. The agile black aggressor responds with a sharp peck in the back. Should the booby continue to resist, the frigatebird grabs its tail, stalling its progress and plucking out a few feathers. Finally the booby submits and regurgitates its fish. The thief peels away to swoop on its falling prize, snatching it from midair in time to arrest its dive just above the water. The long black wings flex once, leaving paired tip ripples on the surface before the wind deflects the raider back up into the sky.

As they have adapted to living exclusively on the wing, the frigatebirds have lost the need for flexible ankles. Their vestigial feet are by now useless for moving the birds through water or over land. The frigates never take a step during their entire lifetime. Once they perch, they are fixed in position until they again take off.

They roost and nest in mangroves—trees that, like the frigates, have come to live not over land but over water. The birds choose a perch tall enough that they can fall away directly into the breeze. The sky above their rookeries is criss-crossed all day by frigatebirds that seem to prefer wheeling through the air to perching in the branches. Adult males show off, inflating the red balloon that expands to fill the space from their throat to the hook at the tip of their bill.

Nests are built from sticks not washed up on the shore but plucked floating from the water. And just as the frigates steal fish gathered by other birds, they may steal nest-building material from their neighbors' unattended nests.

For parents that alight neither on land nor on water, food sources are restricted. So nestling chicks spend much of their time in solitude. Parenting frigatebirds leave their nesting sites to spend weeks at sea, searching for fish they can capture while on the wing, sleeping in the air as they pass the days and nights. The single hatchling in each nest grows slowly; fledgling birds continue to be fed for their entire first year—the longest parental care period of any bird.

The parents fly hundreds of miles away, searching over an ocean that is mostly barren and unpredictable. They overfly hungry boobies who are also searching. They negotiate headwinds that turn the sea into chop. Ultimately, their persistence succeeds. They may find a feeding pod of humpback whales that has chased anchovies up from below; they may dip hatchling sea turtles from the surf beside a desert island; at the center of a stretch of ocean that is otherwise featureless for a thousand square miles, they may spot the glint of silver and scoop up a few lost silversides air breathing through the surface.

The congruence of these seafarers with their environment is completed by their steadfast command of the compass. They navigate the trackless waste, following an internal sense of direction that guides them directly back home after days and weeks of random wandering. At the conclusion of a successful foray, the birds set their sights on that particular point on the distant convergence of sky and sea below which lies their nest. They lock their wings and rise through the air, then glide straight into the distance, diminishing to dark silhouettes that disappear into the haze on the far horizon.

Science Notes

Five very similar species make up the frigatebirds (genus *Fregata*), found worldwide over tropical seas. *F. magnificens* is found over coastal waters around Central and South America (Diamond and Schreibner, 2002). These birds have a distinct silhouette, with an albatross-like aspect ratio (wing length/area) of around twenty, optimal for gliding (Weimerskirch et al., 2003). In contrast, the booby has an aspect ratio of around six, one-third that of the frigatebird. Kleptoparasitic predation on other seabirds accounts for 10–15 percent of the frigatebird diet (Diamond and Schreibner, 2002). Living in the sky over open water for days on end, the frigates are thought to sleep on the wing (Weimerskirch, 2004); they may provide an example of a creature that can alternate hemispheres—sleep one half of its brain at a time. They are mute— the sounds they make at their rookeries may be generated mostly by the rattling together of their bill mandibles.

Euphausiid crustacean krill is a tropical species smaller than arctic krill. The red color, because it absorbs yellow-green light, may help to conceal them from their nocturnal predators after they have attacked and eaten phosphorescent dynoflagellate algae (for further description of predation on phosphorescent dynoflagellates, see Daubert, 2006, chap. 2). Krill congeal into defensive, viscous balls of millions of animals when attacked by herring.

When the krill's numbers explode under optimal growth concentrations, they tint the sea surface an orange pink—hence the name Vermilion Sea in the Gulf of California, where the krill flourish; flotillas of fifty or more magnificent frigatebirds can be seen there. A frigatebird gliding slightly above three hundred feet looks down on a thousand square miles of ocean. Silversides are fish in the family that includes the mullet, the smelts, and the grunion.

References

Daubert, S. 2006. *Threads from the web of life: Stories in natural history.* Nashville: Vanderbilt University Press.

Diamond, A. W., and E. A. Schreibner. 2002. Magnificent frigatebird (*Fregata magnificens*). In A. Poole and F. Gill, eds., *The birds of North America*, no. 601. Philadelphia: Academy of Natural Sciences; Washington, D.C.: American Ornithologists' Union.

Weimerskirch, H. 2004. Wherever the wind may blow. *Natural History* 113:40–46.

Weimerskirch, H., et al. 2003. Frigatebirds ride high on the thermals. *Nature* 421:333–34.

Part 4. Forest

Neither knows of the other. The paths of a rhinoceros beetle (*Eupatorus*) and a darkling beetle (*Tribolium*) cross, unbeknown to either, as they travel the forest floor in search of mates.

The Bitter Taste of Success

A blister beetle has blundered into a spiderweb, and now it waits, suspended in silk, for the spider to come and release it. The beetle dangles upside down, its bulbous wing covers hanging inverted. Their red and black chevrons announce that this insect dares to fly in the face of danger and survive.

The beetles are clumsy and slow, but they are survivors nonetheless. Their members have steadily expanded the beetle niche, which now includes even open territories dominated by aggressive insectivores such as the birds, the mantids, and the orb spiders.

The beetles have followed an unusual pathway to success. They have abandoned defense through speed and evasion, retreating instead behind the protection of an unwieldy shell. That shield created a niche for them that amalgamates the strategies of the two insect archetypes—those that fly and those that crawl.

The beetles can spread their wings and venture as far as the wind may carry them. Then they can stow those wings and dive into thorns, wedge their way beneath sappy splinters of bark, dig into the leaf mold—excursions that would destroy the diaphanous membranes on the backs of other types of flying insect. Beetles have discovered that incorporating the lifestyle of a creature that crawls with that of one that flies provides them access to much more living space than can be occupied using either strategy singly.

When they touch down, the beetles raise their antennae to gauge the local aromas, flavors, and vibrations. They have exploited their combination lifestyle to the fullest by sharpening their chemical senses—perfecting their command of the scents surrounding them and developing the potential of the chemistry within them.

Their rigid body makes them passive, near-sighted, and slow to take wing—characteristics that would seem to preclude their living in the wide-open spaces patrolled by large, visual hunters. But as the beetle species have diversified into more and more habitats, their internal, defensive chemistry has allowed them to expand their range from the concealment of shelter or darkness into the most exposed sun field.

As it twists in midair, the blister beetle struggles ineffectually against the spider's sticky strands. A single thread entangles its spiny back legs, and where they poke against each other, both legs bleed. The silken landing caused no injury, but this beetle is a reflex bleeder—translucent amber droplets bead on the leg joints. The blood contains cantharidin, which blister beetles synthesize in their own cells. The bleeding is a defensive response not detrimental to the insect, serving only as a reminder of the virulence of its blood. Cantharidin is a blister agent. Predators back away when they contact blister-beetle blood. Spiders cut blister beetles out of their webs and drop them if they sense it.

The beetle is too poisonous to eat, so it is protected from those who have experienced its scent. Hungry birds ignore blister beetles; ants are repelled. Cantharidin can be irritating or severely toxic to mammals, depending on the dose. Spanish fly is not a fly but a beetle—a blister beetle—and cantharidin is the active ingredient of potions derived from Spanish fly extract. Its abuse has had debilitating and even lethal consequences time and again over the course of human history.

Their chemical repellency gives otherwise defenseless beetles the capacity to occupy highly visible niches with impunity. Many of the beetles encountered by day are brightly colored. The color is a warning, indicating that the wearer bears a protective chemical defense. The milkweed beetles, either a crimson red or an iridescent blue black, are left unmolested on their sunny feeding grounds—their color is a reminder of their noxious taste. Ladybugs (not bugs but beetles) pursue aphids in their colonies on the exposed ends of shoot tips. Their warning coloration—orange and black like the monarch (who also feeds on milkweed)—advertises their unpalatability to all who notice them.

These day-flying beetles have won places on the list of the hundreds of thousands of beetle success stories. They depend on their toxins for their safe passage and invest their energies synthesizing those defensive chemicals, or recovering them from the plants they feed on, or acquiring them by other, still more inventive means.

Soon, the blister beetle continues on its way, preening off the last shards of spiderweb, resuming a quiet life, senescing a few weeks later and dying a natural death. Then, male members of another family of beetles, the anthicid beetles, will seek out the dead body. Those beetles do not back away when they encounter blister-beetle toxin; on the contrary, they eat it. They are cantharidiphiles—immune to its effects. Their exquisitely cantharidin-sensitive antennae lead them through acres of dead grass back to its dead blister-beetle source. Though they cannot manufacture the toxin themselves, they can acquire it, to their advantage, by eating blister-beetle parts.

The milkweed beetles acquire protection from their predators by consuming and retaining the toxin that protects the milkweed plant from its predators. The anthicid beetles appropriate the protection originally enjoyed by the blister beetles the same way, by consuming and retaining blister-beetle toxin. Male anthicid beetles employ the chemical compound borrowed from the blister beetle not only for their own safety but to further their success with females of their kind.

Female anthicid beetles can also sense the blister-beetle toxin; they look for it when they are approached by a prospective mate. They find males who carry cantharidin to be attractive, and, conversely, they reject males who do not carry enough of it. The toxin is key to reproductive success in both sexes. Females prefer to mate with males perfumed with poison; those females receive a dose of cantharidin along with the sperm—males transfer to them a protective agent that renders them unpalatable to predators.

After receiving the toxin, the female transfers it again, passing some along to her eggs. So, just as blister-beetle eggs are protected from predators by their cantharidin toxicity, so are the eggs of the anthicid beetles. Unhatched anthicid-beetle larvae are protected from still other kinds of beetles—those that prospect through the soil to prey on eggs. These anthicid larvae profit from the dissuasive blister-beetle toxin they have inherited from their mother, who received it from her mate, who acquired it by feeding from the primary blister-beetle source.

Fireflies are not flies but beetles. They are active after dark, but the night does not conceal them—firefly beetles give away their locations while making themselves visible to potential mates.

There are hundreds of species of firefly beetles. Each emits a characteristic series of long and short flashes—a specific code by which members of particular firefly species can recognize each other through the darkness. To overcome the disadvantage of their high visibility, many of the firefly beetles protect themselves by producing a toxin that renders them unpalatable, just as high-visibility day-flying beetles do. Other firefly beetles do not produce the toxin but acquire it by subterfuge.

The females of one non-toxin-producing firefly-beetle species acquire the protective toxin not by feeding on dead members of other toxin-producing species, as the anthicids do, but by feeding on live males of toxin-producing species of firefly beetle. Those females lure the males of the other species of firefly to their deaths by altering the sequence of flashes they send to match the code of that species.

The toxin they acquire through their predatory diet protects them just as if they were toxin producers themselves. These protected females then use the toxin to further their own reproductive success. They transfer some of it to their eggs, thus conferring the protective chemistry of a different firefly species on their own unhatched young. And to remind predators on the ground that these are the eggs of toxic fireflies, the eggs glow in the dark—with the luminous color of a firefly.

Their mastery of the chemical cues around them, and their control of the chemistry within them, allows the highly visible beetles to safely occupy a vast living space that lies in plain sight. This use of chemistry is just one example among many of the proclivity of the beetles to spread out into more niches than does any other order of animals on Earth. The beetles are the most diverse order in the most successful class of animals on the planet—the insects. They are the largest order, by number of species, in the animal kingdom. A random sampling of eight animals from the face of the earth would contain five insects, and two of those would be beetles. There are more known species of beetles than of plants. And there are proportionately still more of them waiting to be discovered.

Most of what we have discovered about the beetles is what is in plain sight: their shapes and colors, their ranges, their behaviors. We still understand relatively little about the invisible molecules they produce or respond to—the chemical stimuli that motivate them and provide for their defense. The beauty of their evolutionary accomplishments is in their chemical ingenuity, about which we have essentially everything yet to learn.

Science Notes

The beetles combine defensive armament with flight to their great advantage (Crowson, 1981), giving rise to greater diversity in their order of insects than is found in any other (Hunt et al., 2007). Defensive chemistry (King and Meinwald, 1996) has become a factor in the selection of the niches many of them can occupy. The survival value of their defensive chemistry (Smedley et al., 1996) is often advertised in their aposematic (warning) coloration or appearance. The beetles balance the selective advantage of an altered chemistry with the metabolic cost of the production of their defensive chemicals. Some beetles avoid the cost of production by acquiring defensive chemistry through the ingestion of other beetles (Eisner et al., 1996, 1997).

References

Crowson, R. A. 1981. *The biology of the Coleoptera*. London: Academic Press.

Eisner T., et al. 1996. Chemical basis of courtship in a beetle (Neopyrochroa flabellata). *Proceedings of the National Academy of Sciences, U.S.A.* 93: 6494–6503.

Eisner, T., et al. 1997. Firefly femmes fatales acquire defensive steroids (lucibufagins) from their firefly prey. *Proceedings of the National Academy of Sciences, U.S.A.* 94:9723–28.

Hunt, T., et al. 2007. A comprehensive phylogeny of beetles reveals the evolutionary origin of a superradiation. *Science* 318:1913–16.

King, A., and J. Meinwald. 1996. Review of the defensive chemistry of coccinellids. *Chemical Review* 96:1105–22.

Smedley et al. 1996. Predatory response of spiders to blister beetles (family Meloidae). *Zoology* 99:211–17.

Fair-Weather Desert

The yellow warbler found light enough to start her day before dawn—when the branches around her first appeared in silhouette. The early glow was weak, but brighter than the starlight she traveled by in the dead of night across some terrains. She was rested and ready to keep moving—through the land of late summer drought around her. So she dove from her roost and set off, while the last stars still glimmered in the west.

The dusk allowed her to travel through the trees with some sense of concealment. After sunup, her canary yellow form drew attention—leading her to move quickly so as not to present a stationary target for her predators. But constant motion left her in constant hunger. She was a lightweight—a high-metabolism creature who did not carry much stored energy—and she had to find food as she went. But since she had stayed with her nest too long in the boreal woods to the north, then started her migration later than was safe—her window in time was closing.

The stress of constant motion during migration, and constant care of her young during the nesting season, contributed to the high attrition rate for her kind. A few weeks and a thousand miles ago she had fledged the last of four chicks from consecutive broods. She and her mate had been reproducing at a rate that could double the yellow warbler population every year, if all the young survived—but most would not. Yet she put herself at risk with the effort to give them the chance. She would pursue that strategy as long as she could. Species that do not overreproduce in this way have long since disappeared from the Earth.

Early northern storms had pursued her through the first stage of her migration, leaving no time to forage and regain her weight. Now, in the dryer reaches of her trip, insect prey was scarce. And this day would offer even more challenges. Soon she would come upon a series of river valleys that had once been fair-weather havens—cool rest stops that promised food and shelter to sustain her. Now they offered more hazards instead.

From her distant vantage in the foothills, the first of those valleys spread out below. The land looked as if it had been buried by avalanches of debris washed down from the adjacent mountains—leaving the trees struggling to recover. The rich, dark forest and richer riparian zone in the bottomlands where the river should have run were gone. On previous migrations she had discovered that valley bottoms farther along her path were growing to look similar to this one.

She paused in the open branches of a ghost pine, scanning the altered landscape ahead. Finally she dashed across the stretch of broken soil and weeds at the boundary of the area, leaving behind her familiar oaks, pines, and willows. One benefit arose when entering this aberrant world—hawks that preyed on small birds hesitated to go there.

The transformed land was covered by a maze of angular blocks fitted close against one another. Trees and shrub species she did not recognize rose in the notches between them. It was a fair-weather day, but the area was dry and dusty from a bird's-eye-view—a desert where rainwater ran across impervious surfaces and disappeared down holes, and where the few streams ran mostly underground. Most of the plants were stressed from lack of water, and the weakened foliage was infested with mites, scale insects, or aphids—so there was some forage for the warbler to find among the branches she moved through.

She stayed as high in the tree cover as she could. On an earlier migration through here she had spied a pool of freshwater suspended on a pedestal low in the understory. It was deep enough to bathe in, but as she descended to its edge she noticed a feline shadow poised in ambush in the ferns.

She had twisted her path barely in time to escape the claws and teeth that flew at her. The small cats that inhabited the area were stealth hunters by day, territorial fighters against each other by night. Their young were insectivores, but the adults killed birds as often as they could, even though they left their kills uneaten. These predators stalked every inch of this alien habitat. Their territories ended abruptly at its boundaries, where they became prey for coyotes.

There was plenty of forage at ground level if the warbler wanted to risk her life to hunt there. Weeds that should have been browsed to extinction by rabbits and deer were profuse, providing habitat for ants, slugs and snails, cutworms, earwigs, and pill bugs. These crawlers reproduced here as fast as they could. In the woods and fields they were kept in check by birds that foraged and nested on the ground—quail, rails, thrashers—but those birds had been eradicated by the felines that now lived here. The uncontrolled hordes of invertebrates quickly destroyed seedlings of the native flora that germinated anywhere within the area.

The warbler could not forage long in the trees, out of concern for the jays and mockingbirds. They were the primary native birds adapted to this nonnative habitat, but the ones that lived here were much edgier than the members of their species that lived in the surrounding hills and forests. In this habitat, the jays wanted to destroy smaller birds. They dove at warblers that entered their territories, using their broader wingspans to pursue and chase them off.

Should the evening overtake her as she crossed this fair-weather desert, the warbler would not rest at ease. This environment remained light all night—most of the stars never appeared through the glowing air. She sat with eyes half open in the shadows of the extended dusk and listened to noises that persisted through the darkness. She was kept alert by arboreal rats prowling in the trees—animals that were unknown along the rest of her route and would be extirpated by owls and foxes should they appear there.

The warbler found little incentive to tarry—even the air itself was irritating, bothering her breathing and stinging her eyes. She usually chose to forgo sleep and keep on moving through such areas—the sooner to leave them behind.

As she approached the far boundary of the transformed valley, she saw some of the inhabitants of the blocky structures that covered this landscape. They were large animals—humans—who lived within the angular chambers. This whole area was a colony of humans, growing larger all the time. On her way out, she passed again across the weedy, bare-ground region—the broken-soil border that surrounded the colony and marked the early stages of its further expansion.

The inhabitants inside their chambers, like all the other creatures, were growing the numbers of their own species as fast as they could. But, unlike other species, their growing populations were not checked by predation, by limits to their food sources, or by the complete occupancy of their niche. They manufactured their own preferred habitat as they expanded their range, without consideration of the consequences their multiplication would have for other species and for the fate of their own future generations.

A month later the warbler sat in the mistletoe sipping rainwater from bromeliads. She was finally at rest in the forest of broadleaf evergreens where she would spend the winter. These tropical surroundings echoed with the calls of birds, insects, and frogs—a chorus restful to her ears. From her vantage point high in the trees she could see a world unknown to other eyes. With receptors on her retina for four different colors—more than most other animals—she could see subtle hues that would lead her to bugs and berries, and to opportunities for shelter unnoticed by competing classes of creatures.

Through the branches she watched a giant python resting motionless along a descending stony edge on the forest floor far below. The snake was a distraction for her—day after day it never moved from its position—yet whenever she ventured near enough to assess the possible threat, it seemed to melt away into a pile of rocks. She could see the snake, always in clear view, only from a great height.

She fluttered closer, detecting no further signs of snakes and noticing no caution in the bearing of other creatures in the area. Finally she forgot about the disappearing serpent and went on to forage through the under-brush. She flitted among the vine-tangled limestone blocks and chased a small moth between the giant fangs, finally perching to sing on the lid of a stone snake-eye as wide as she was tall.

This reach of bottomland was also once a broad, expanding human colony. Warblers from generations long past had skirted its sprawling campus, avoiding the stone temples carved with images of the god-animals of the forests and the low, weedy areas of broken soil at the expanding margins of the civilization.

As with other species, the human inhabitants of these colonies had been multiplying their numbers as fast as they could. They did not realize they had pressed beyond their natural limits when local incidents of parasite infestation and microbial disease became epidemics—spreading one after another through the populace. They did not notice that their agriculture had extended past a similar limit when the blight that once darkened an occasional ear of maize inexplicably expanded, causing devastating crop failures across Central America.

As their growing needs drove them to forage ever farther, these people encroached on the territories of neighboring human populations. Then leaders appeared in each colony to marshal the men into armies to defend their lands and attack the adjacent cities. Overpopulation stresses—expressed through war, disease, and famine—eventually crushed them all, overwhelming their science and civil planning, leaving them bewildered and helpless. Their social structure faltered, but they had forgotten how to live off the land. They watched their cultures die—their cities deserted so quickly that the peak of their grandeur still lived in the memories of their last survivors.

The warbler hurled herself from her perch into the night. She had barely avoided the strike of an arboreal viper who could see her outline plainly in the dark. She threw her wings up and descended feet first, clattering through the leaves to land roughly on a branch she could feel but not see.

There she stood in silence—a waif in the black, overcast night, listening for danger and waiting for first light.

It was a situation she knew well. Warblers would always have their numbers controlled by forces above them on the food chain. They never found themselves competing with others of their own kind for living space.

As they migrated across the ages over the Americas, the warblers had borne witness to other, larger animals that had mastered the challenges of their predators, parasites, and environmental extremes—enabling them to expand their numbers continually. But even when their prosperity was ensured, these climax species did not alter their inherent drives but continued to multiply as fast as they could. The horses, then the bison, then the elk multiplied and peaked; then their overpopulations crashed, leaving only a trace of their past numbers, if that. The warblers had never seen a dominant species that mastered the challenges of its environment, then modified its inherent drives and held its population to a moderate, fair-weather level to perpetuate the days of its prosperity.

Science Notes

Direct or indirect change in the natural landscape wrought by the imposition of colonies of humans is near total (Johnson and Klemens, 2005). The biological diversity in areas submerged by suburbs is severely reduced. Wild species that remain in place become peridomesticated (adjusted to human proximity), including synanthropic (benefiting from human proximity) species, such as jays and mockingbirds. A small suite of introduced, adaptable species of plants and animals—encountered more and more frequently in urbanized areas around the world—comes to the fore. House cats are suppressed in the wild drylands by coyotes, but where they are prevalent (around dwellings), they suppress the native birds (Crooks and Soule, 1999); they are estimated to kill 100 million small birds a year. Among those is the yellow warbler *Dendroica petechia* (Lowther et al., 1999).

Overproduction of young is a standard strategy in almost all animals. Individuals produced in excess of the replacement number for their parents usually perish before they reach reproductive age. But when boom conditions arise and their habitat temporarily supports a larger population, members of the overproduced population survive—until more typical conditions return to reduce the population back to its carrying-capacity level. When that inevitable reduction occurs, the overproducers are overrepresented among the survivors. Over these cycles of boom and bust, nonoverproducers are gradually eliminated from the population.

Birds see their surroundings with tetrachromatic vision (Church et al., 2001), enhancing the hues they perceive and extending their visible spectrum beyond that experienced by other animals.

The migratory warblers were present in Central America when the classic Maya flourished in the first millennium A.D. The fall of Mayan civilization began with the abandonment of the great city of Tikal in 900 and was completed with the desertion of Chichen Itza around 1200. Not enough evidence has survived the jungles for us to know with certainty the factors that brought about this collapse. The stresses of continual wars likely contributed, as may have drought stresses on the overextended farming systems (Kohler et al., 2005). A blight on their agriculture has been suggested (Brewbaker, 1980). The rising global population curve of the human species shows no sign of moderation, despite human "intelligence." The human population growth curve shares the characteristics of those of other animals that have discovered a suitable habitat, then exploited it by multiplying up to and beyond the sustainable limit.

When a population rises in prominence, it becomes a resource to exploit, thereby generating its own new brace of parasites and predators. Along with Pleistocene environmental changes, another stress that may have hurried the extinction of the ancient bison (*Bison antiquus*) or the western horse (*Equus occidentalis*) or the once great herds of elk was the appearance of predatory humans.

References

Brewbaker, J. 1980. Disease of maize in the wet lowland tropics and the collapse of the Maya civilization. *Economical Botany* 33:101–18.

Church, S. C., et al. 2001. Avian ultraviolet vision and frequency-dependent seed preferences. *Journal of Experimental Biology* 204:2491–98.

Crooks, K. R., and M. E. Soule. 1999. Mesopredator release and avifaunal extinctions in a fragmented system. *Nature* 400:563–66.

Johnson, E. A., and M. W. Klemens. 2005. The impacts of sprawl on biodiversity. In E. A. Johnson and M. W. Klemens, eds., *Nature in fragments: The legacy of sprawl*. New York: Columbia University Press.

Kohler, T., et al. 2005. Simulating ancient societies. *Scientific American* 293:76–82.

Lowther, P. E., et al. 1999. Yellow warbler (*Dendroica petechia*). In A. Poole and F. Gill., eds., *The birds of North America*, no. 454. Philadelphia: Academy of Natural Sciences; Washington, D.C.: American Ornithologists' Union.

Tree-Squirrel Fungus

The world's predominant green province darkens broad reaches of the northern half of the northern hemisphere. A great forest there covers nearly twice the land area that the tropical forests cover in the lower latitudes. Northern conifers blanket much of Eurasia in Scots pine, Siberian spruce, and larches, while across the Bering Strait in North America, black spruce, white spruce, tamarack, and aspen grade into white pine and western hemlock. Hardwood trees mix with pines farther down the mountainous spines of western Canada and into the United States.

This vast, vital arm of the biosphere has anchored itself in a circumpolar territory that routinely experiences weather of ice-age extremes. The world's deepest winters leave the soils there perpetually soaked in snowmelt runoff. The constant running groundwater combines with the acidity of decomposing cones and needles to leach the mineral nutrients out of the earth, making this a challenging place for plants to grow.

But the great conifers have conquered that challenge by entering into a complex, cooperative network with a broad assortment of other creatures. This assemblage includes many of the animals and fungi that live in the shadows beneath the soaring crowns. Great and small alike are part of a mutual commerce in sap and blood carried in the veins and the guts of organisms that insure each other's survival.

This pyramid of cooperation stands on a foundation woven of glassine fungal strands in the soil. Those microscopic tendrils colonize the roots of the trees, extending farther through the soil than the distance between the great trunks to underlap every inch of forest floor—their fine filaments fusing with one another to connect every tree with every other. Through them, all the plants in the forest are linked below ground into a single vast, symbiotic superorganism.

If they have access to enough water and nutrients, the trees will never cease their growth—they do not reach maturity at a specified age and then stop but grow ever fuller. Their demands on the resources at their bases increase along with their ages and sizes, until they exhaust the essential elements within their reach. Starvation for any one of these elements leads to senescence and death.

But while in their prime, the trees make copious amounts of sugar out of sunshine and thin air—more than enough to fuel their growth. The extra sugar is tapped by the symbiotic root-associated fungi. In return for the sustenance, those fungi probe far and wide throughout the soil to capture water and the scarce vital salts dissolved in it—phosphate, potassium, nitrates, and other trace minerals. The fungi feed those nutrients back to the trees to sustain the barter relationship upon which the forest depends.

Newly germinated pine and spruce seedlings eventually capture and recycle the store of minerals accumulated by earlier generations of trees. Those nutrients are released back into the soil fungi that infiltrate fallen trunks and rot the wood. But the new seedlings awaken to a far more vigorous pursuit of their destiny after they have made contact with the synergic fungi that colonize their roots.

One of the primary components of this cohort of symbiotic root fungi is the truffle. Its subterranean network of invisible threads is welcomed by the trees' roots—while unrecognized fungi (which may be pathogenic) encounter a rebuff from the trees' defenses. The truffle fungi on which the trees depend are themselves dependent—during one critical phase of their cycle—on one of the animals of the forest. Though they propagate by spores, the truffles cannot disperse their spawn on the wind, unlike the many fungi that float weightless spores from beneath caps or conks pushed out into the air through soil or bark. No such levitating breezes blow where the truffles live; no bees visit underground, nor do butterflies or bats. The truffle remains buried during its entire life cycle.

To entice someone to assist the spread of their spores, the truffles' subterranean threads congeal during the summer—to metamorphose into one of the most edible delicacies of the wood. They produce seductively scented and richly flavored fruits—a choice food, if someone would just dig them up.

That someone would be the northern flying squirrel. These foragers locate and feed upon more than twenty species of buried truffle fruit. They are nocturnal animals with big eyes and flat tails, with soft silver-gray fur below, cinnamon brown above. The broad, furred flight membrane—the patagium, attached from wrists to ankles—renders them clumsy when they come down to earth to dig. But the squirrel does not stay down for long.

The fruiting of the truffles is unpredictable, prolific in one glen or during one autumn, meager the next. Flying squirrels must search far and wide each night for that chance particular bouquet of ripeness lingering just beneath the ferns and sorrel. These squirrels patrol a much broader territory than do the chipmunks, ground squirrels, and tree squirrels that feed on pine nuts. A flying squirrel's territory can stretch for twenty acres. To sample such an area every night, the creatures must move faster than tiny feet could run along the ground.

The squirrel visits its dispersed subterranean gardens by casting off from the highest branches, its legs spread-eagled, and waiting for the wind of its descent to catch beneath its patagium. Then it can glide from tree to tree to reach its destinations. Descending trunks it has chosen as most promising and least risky, the squirrel forages briefly, then climbs back up to resume its lofty commute. Its ground track covers miles of forest floor every night, but the flying squirrel covers it by air, not on foot.

The squirrel pursues a zigzag flight path, avoiding directly crossing open spaces. The diminutive animal prefers to stay among branches that obstruct the sight lines and flight lanes of its predators. It may need to dive for cover the instant it senses the approach of a winged nemesis.

The truffles might advance their transparent strands six millimeters in a single spring day's growth. But they advance their spores much farther by producing them within their sought-after fruit. A flying squirrel can carry fungal spores in its gut 60 feet in a single bound, after springing into space from a pine promontory 150 feet up. This association with the squirrel gives the fungus an advantage in its competition with other, less savory root-associated fungi. Their competition with each other for the squirrel's attention continually drives the various truffles each to be more delectable than the next.

Should the flying squirrel confront danger on the ground, the membrane that floats it through the trees becomes a hindrance, restraining the little mammal's escape. It cannot muster the instant dash of a ground squirrel or chipmunk, so this squirrel descends only under cover of darkness. It rests on high by day, relying on another fungus to conceal it from its predators.

The fungus that provides the squirrels' shelter is witches' broom. In its turn, it is dependent on the trees—completing the triangle of interdependency among the trees, squirrels, and fungi. Witches' broom's parasitic lifestyle causes the disorganization of the growth patterns of conifer foliage. New shoots proliferate on infected branches in an aberrant profusion so dense that they shade and stunt each other, twisting into mutant tangles. Tight thickets of stems result, sprouting vertically from horizontal branches of spruce or fir at random places throughout the canopy. These knots in the foliage provide secure den sites in which the flying squirrels knit for themselves a blanket of lichen and moss to sleep away the day.

Later in the season the witches' brooms produce blooms of spores that coat fir needles and fur coats alike in rusty-orange spore powder. Should it not groom itself meticulously, the flying squirrel spreads the witches' broom infection as it pursues its rounds through the trees.

As the grand trees expand above the top of the forest canopy, their need for minerals finally overtakes the soil's ability to provide. Their root systems expand, crowding into the roots of adjacent trees with the same needs. At that stage the trees enter into another symbiosis with a different set of organisms. The keystone member of this guild is yet another forest fungus—the causal agent of heart rot.

As the decades pass, the trees are buffeted by the impact of other trees falling around them. One of the primary functions of their stout trunks is to resist the strikes of tons of falling timber that occasionally crash against them. If every year were a second, then several times a minute the growing hemlock or fir or redwood would shudder at a glancing blow delivered by one of its older, dying neighbors toppling from a base one tree-height radius away. These impacts regularly rake each standing tree; they take out divots from the bark or shear off branches to expose the cores in the branchwood briefly before the tree closes its wounds.

The spores of the heart-rot fungi are inoculated into standing trunks through such breaches. Heart rot propagates down the central axes of the largest boles, advancing only inches a year. It lives on the dead wood at the radial center of the trunk, surrounded by hundreds of growth rings in the mature trees it prefers.

The fungus is a benign tenant in the tallest towers of the forest. All the oldest wood supports it. It does not harm a tree's living parts, avoiding the vital layer of sapwood just beneath the bark. This outermost cylinder of cambium constantly enlarges itself outward from the central axis, leaving behind the previous season's growth in the form of concentric tubes of inert, dead heartwood. Should a spur of heart-rot decay come in contact with the sapwood, fungal propagation halts.

As the fungus eats the heart out of the great trunks, it liberates the essential minerals sequestered in the otherwise inert core wood, making them available for resorption by the trees. Hollowing of the core has little impact on trunk strength, just as a pipe is structurally similar in strength to a solid rod made of the same material. And the decomposition of the heartwood lightens the weight that the standing cylinder must bear. Once the heart-rot phase of a tree's growth is initiated, a second set of creatures furthers the symbiosis as it invades the rotting spaces—these are beetles. Many species, from smaller click beetles to the great, ornate stag beetles, spend their larval stages in the moist, crumbling matrix of fungal rot that was once solid heartwood. Over the years, they follow the leading edge of the fungal rot through the tree's core, opening the space into great long galleries.

Then the bats, tree swallows, and owls invade those gallery spaces in search of resting and nesting sites, their entry facilitated by another family of birds that depends on the fungus to soften the heartwood—the wood-peckers. When they are on the wing, away from their tree-cavity roosts, foraging bats, swallows, woodpeckers, and smaller owls all moderate the burden of the herbivorous insects upon the forest.

This succession of roosting and nesting species occupies the hollow cores of the oldest of the forest monoliths as the space becomes available. Flying squirrels bear their naked, helpless pups in the shelter of nesting cavities in tree holes. Born with eyes and ears fused closed, these pups are more defenseless than newly hatched warbler chicks. They will not fly until five weeks after birth.

All these creatures—beetles and owls, bats and squirrels—share space in the vertical caverns of the older forest trees. In addition to the benefits the trees realize from control of insect outbreaks by their birds and bats, the trees reap a second windfall from their tenants. The deepening hollow spaces become a repository for the accumulated droppings of all the tree's inhabitants. Nitrogen-rich excrement, the end product of a harvest from miles around in every direction by the resident animals, is deposited in the central cavities of these trees. This guano will eventually be broken down by the action of other fungi and microbes into a form that the trees can absorb from within to augment the nutrition they search for in the soil.

The additional nutriment further extends the tree's lease on life. The forest giants at the hollow-gallery stage continue to support their fungal burden and grow on for centuries after they first contract heart rot. Douglas fir can live beyond a thousand years; Alaskan yellow cedar—three thousand. It is impossible for us to say just how far the harvest of nutrients imported by their tenants extends the great ages of the residents of pristine old-growth forests—because the growth-ring records of their years have been obliterated by the expansion of the heart rot at the center. But almost all the sentinel trees in a silent, mature grove support extensive rot, and the colonists that inhabit it.

In the highest of these trees, sitting highest on the food chain, is the spotted owl. This predator lives deep in the heart of the northern forest, where the hemlock and Douglas fir loom hundreds of feet over stream-cut glens.

The bird represents a pinnacle of predator adaptation to the forest niche. Its feathers are modified for stealth—each feather carries a fine fringe on its leading edge so that, in contrast to the feathers of day-flying birds, owl feathers don't sigh in flight—their approach through the night is silent. The owl's broad wings stretch four feet from tip to tip, ten times the wingspan of the flying squirrels it preys upon.

Flying squirrels make up the major part of the northern spotted owl's diet. The owl's chosen habitat overlaps with the range most preferred by its prey. Both are found in those districts where the northern forest is at the peak of its productivity. A healthy spotted-owl population reflects a forest in its most biodiverse and resilient form.

The network of animals and microbes that supports the northern woods is but one community among the many interdependent life zones that occupy the world's continents and seas. These zones feed upon each other to stabilize the biosphere, with the northern forests anchoring the equilibrium. Global hydrology is modulated in the North, where the woods affect the sky's water vapor burden through the biological buffer of their transpiration. The verdant, subarctic mantle suppresses reflection back into space of the light and heat falling most directly on the world at their latitude in high summer. The influence of those northern woods spreads beyond their farthest southern reach as they moderate cycles of seasonal rainfall and evaporation throughout the hemisphere.

The northern forest moderates the world's air temperature by storing atmospheric carbon dioxide within its branches and in the resinous humus below them. The great pinewoods remove kilotons of atmospheric carbon dioxide per acre as the trees mature each year.

The threads that link the interdependent members of the northern woods community are fragile. The diversity and number of symbiotic fungi are diminished by disturbance in the forest. In the absence of the truffles and their soil-fungus allies, seedling pines fail to establish, and stress in the roots of the older trees threatens their survival in a summer's drought, in a severe winter wind, or during an onslaught of disease or insect pestilence. Stressed woodlots are sources of boring-beetle outbreaks and infestations of pathogenic fungi that can threaten healthy trees over wide areas.

We cannot easily judge the health of a forest by looking at the trees. The indicators of their long-term prosperity are not apparent during a walk beneath the towering arches—the symbiotic triangle of fungi and squirrels working with the roots operates in the dark and underground. Witches' broom, heart rot, and the other fungal members of the forest work out of sight.

The owls, however, are more easily monitored. We can see them in their trees or count them by their calls, which carry through the branches for hundreds of yards on still, foggy nights. We can monitor the variation in their numbers from year to year. Through them, we can infer the success of all the other components of a food chain on top of which they ride. The owls are a barometer through which we can confirm that our interaction with the forest has become sustainable. A stable northern tier to our biosphere would ensure that our own long-term tenure here will be stable as well.

Science Notes

The forest is an interwoven, interdependent tapestry of threads, linking the trees with fungi, insects, birds, and mammals into a network that fosters the survival of all the participants, as well as that of other creatures not directly part of the weave. The entire biosphere benefits from the vitality of the boreal forests through their moderating impact on the atmosphere (Snyder et al., 2004); that moderating influence can be diminished by anthropomorphic influences such as logging, and by insect outbreaks (Bond-Lamberty et al., 2007). The boreal forest has been calculated to sequester two to five hundred gigatons of carbon; the temperate (nonfreezing) coastal rainforest, as much as five hundred to two thousand metric tons per hectare (Rayment and Jarvis, 2000). That exceeds the amounts sequestered in tropical forests. Eleven percent of Earth's terrestrial surface is shaded by the northern forests, compared with about 7 percent for the tropical forests. The northern forests receive 50–60 inches of rainfall per year, and up to 140 inches per year where the trees reach their climax diversity along the coast of the eastern Pacific; there it is a rainforest, in the tropical sense, with more biomass per hectare than the Amazonian average. The boreal forest's strong absorption of heat and light and its low evapotranspiration result in a rapid spring warming response to increased day length. The differential there between conditions near ground and the air aloft alters the convective boundary layer, impacting global cloud formation and precipitation.

Trees supply us with oxygen through the photosynthesis reaction that converts airborne gasses to solid fiber. That reaction combines carbon dioxide and water from the air to produce carbohydrate (sugar), with oxygen as a byproduct. That reaction is written: $6\,CO_2 + 6\,H_2O \rightarrow C_6H_{12}O_6 + 6\,O_2$. Plants polymerize the sugar into cellulose to produce the structural parts of leaves, wood, et cetera.

The fungi are pivotal in the maintenance of the forest (Rayner, 1993). Few if any plants exist in natural ecosystems independent of symbiotic associations with fungi (Petrini, 1986). Fungi play a central role in the health and structure of plant communities (Read, 1999). *Tuber gibbosum*, the Oregon white truffle, associates with Douglas fir (*Pseudotsuga menziesii*), as do many of the other species of north-woods truffle. The truffles have long been known for the enticing complexity of their bouquet. The attraction of truffles has been described, for example, for *T. gibbosum*, as smelling of roasted hazelnuts, carrying the flavors of fine cheeses and tropical fruits, and leaving the aftertaste of a dry white wine. On the human palate, the truffle is the quintessential wild mushroom delight. Flying squirrels have been observed to prefer truffles as well, choosing them ahead of, say, the tannic acid astringency of acorns or the gustatory challenge of a spiny cricket—truffles make up most of the diet of the northern flying squirrel.

Witches' broom is in the rust family of fungi. Rusts are named for the color of the spore bloom. Brooming appears to involve an endocrine derangement of host plant growth and is caused by a number of different parasites on a wide variety of plants. Rust fungi pursue an obligate cycle of alternate hosts: for example, fir broom spends part of its life cycle on chickweed; spruce broom is found only in areas where a species of manzanita (bearberry) also grows. A nonfungal causal agent of broom in western hemlock, Douglas fir, and Ponderosa pine is dwarf mistletoe.

Though loggers may refer to the great trees that support heart rot as "overmature" or "cull," there is an abundance of heart rot in old-growth forests (Hennon and DeMars, 1997). Nitrogenase activity in the cavity spaces produced by heart rot (Harvey et al., 1989) reflects the abundance of minerals there, made available to the trees by the rot. The actual growth support conferred on the trees at the hollow gallery stage by their resorption of the guano imported by their tenants has not been well quantitated. In theory, trees grow but do not age; their lives would never end if they did not run out of nutrients to support their ever-increasing demands.

Along with the keystone fungi, woodpeckers are also keystone species in the colonization of the forest spaces (Daily et al., 1993); they may nest exclusively in older trees that support heart rot (Connor et al., 1976; Jackson, 1977). The northern spotted owl (*Strix occidentalis*) lives in the northwestern coastal margin of North America (Gutierrez et al., 1995), sharing the temperate rainforest—the most productive section of the northern woods—with the flying squirrel (*Glaucomys sabrinus*). The squirrels do not generate lift as they fly; they glide, or volplane. Other races of spotted owl live to the south, beyond the squirrel's range. Monitoring programs for spotted owls (Courtney et al., 2004) and for the diversity of all species in districts of the north woods allow us to gauge the success of our efforts to interact with the forest sustainably.

Remnants of the climax north-woods forest at its most diverse and resilient are rare. The steadily increasing human population is having a widening impact on the biosphere we depend on. Human activities now appropriate a third or more of the total productivity of the global ecosystem (cited in Foley et al., 2005). Anthropogenic increases in the temperature of the atmosphere, pollution of the rains, physical destruction of sections of the forest through urban expansion or road building or unsustainable timbering—all can sever the threads of interdependency on which the forest depends for its vitality. Disturbed, regrowing forest, in which the mix of tree species is out of balance with that established over the millennia, contains lower numbers and diversities of root-associated fungi (Read and Birch, 1988). These trees are more susceptible to infestations of beetles, to drought stress and forest fires in late summer, and to wind damage in winter. Degradation of the interaction of the northern forest with the biosphere will only magnify the other problematic consequences of humankind's excesses.

References

Bond-Lamberty, B., et al. 2007. Fire as the dominant driver of central Canadian boreal forest carbon balance. *Nature* 450:89–92.

Connor, R. N., et al. 1976. Woodpecker dependence on trees infected by fungal heart rot. *Wilson Bulletin* 88:575–81.

Courtney, S. P., et al. 2004. *Scientific evaluation of the status of the northern spotted owl.* Portland, Ore.: Sustainable Ecosystems Institute.

Daily, G. C., et al. 1993. Double keystone bird in a keystone species complex. *Proceedings of the National Academy of Sciences, U.S.A.* 90:592–94.

Foley, J. A., et al. 2005. Global consequences of land use. *Science* 309:570–74.

Gutierrez, R. J., et al. 1995. The spotted owl (*Strix occidentalis*). In A. Poole and F. Gill, eds., *The birds of North America*, no. 179. Philadelphia: Academy of Natural Sciences; Washington, D.C.: American Ornithologists' Union.

Harvey, A. E., et al. 1989. Nitrogenase activity associated with decayed wood in living north Idaho conifers. *Mycologia* 81:765–71.

Hennon, P. E., and D. J. DeMars. 1997. Development of wood decay in wounded western hemlock and Sitka spruce in southeast Alaska. *Canadian Journal of Forest Research* 27:1971–78.

Jackson, J. A. 1977. Red-cockaded woodpeckers and pine red heart disease. *Auk* 94:160–63.

Petrini, O. 1986. Taxonomy of the endophytic fungi of aerial plant tissues. In N. J. J. Fokkema and J. van der Heuvel, eds., *Microbiology of the phyllosphere.* Cambridge: Cambridge University Press.

Rayment, M. B., and P. G. Jarvis. 2000. Temporal and spatial variation of soil CO_2 efflux in a Canadian boreal forest. *Soil Biology and Biochemistry* 32:35–45.

Rayner, A. D. M. 1993. The fundamental importance of fungi in woodlands. *British Wildlife* 4:205–15.

Read, D. J. 1999. Mycorrhiza: The state of the art. In A. Varma and B. Hock, eds., *Mycorrhiza*. Berlin: Springer-Verlag.

Read, D. J., and C. D. P. Birch. 1988. Effects and implications of disturbance of mycorrhizal mycelial systems. *Proceedings of the Royal Society of Edinburgh* 94:13–24.

Snyder, P. K., et al. 2004. Evaluating the influence of different vegetation biomes on the global climate. *Climate Dynamics* 23:279–302.

Focal Point

Two birds approach each other along the chill, shaded bole of a great old Douglas fir—one walking straight up, one walking straight down. The white-breasted nuthatch grapples noisily with the bark, arresting his headlong plunge with every step down the cylindrical surface. His talons and upturned bill dislodge a small shower of duff and bark chips as he goes. He watches these particles float away before him, his quizzical eyes hidden in black stripes set off from his black cap by white eyebrow lines.

The brown creeper is less obvious—camouflaged in dark, stippled plumage—quietly making her way upward. With her stiff tail pressed against the bark behind her, she scuttles from side to side, avoiding the largest of the particles falling from above. Then she stops to search for food—the spiders, mites, and pupae that can best be seen from underneath. From his opposing perspective, the nuthatch encounters a menu of different insect fare—prey species with alternate concealment strategies. Thus, the two birds do not compete directly but coexist in the vertical dimension they share.

They pass each other on opposite sides of what appears to be a slim, raised flake of bark. Each bird glances at the low wedge but then quickly looks over it to regard the other bird. They pass closer than either would allow a member of its own species to approach, then continue on their ways—one up, one down.

For an instant, both birds had considered whether this raised flake might be breakfast. It was textured in the concentric whorls of annual growth layers in bark—a pattern commonly employed as camouflage by moths and other insects that rest exposed on tree trunks by day. First light in the boreal forest finds these insects paralyzed by the cold, unable to fly, dependent for their survival on escaping notice behind their mock-wood texturing.

But neither bird had observed the aerodynamic curve of a lepidopteran wing—there was none there to be seen. Instead, each bird's quick inspection had revealed a silver crescent standing out clearly on the irregular dark background. Their eyes were drawn to this shiny focal point, which had the reflective appearance of a slime mold or other viscid, shapeless white fun-

White C. A shiny mark punctuates the cryptic
coloration of the anglewing's underwing.

gus conforming to the bottom of a recess in dead bark. Such wet, inedible spots are commonly encountered by these birds, who have become inured to the sight of them. Both foragers quickly scanned the white shape standing out against the darker surface, and then their searches moved on.

As the two prospectors diverge, their roles invert. Anything dislodged by the creeper would now fall on the nuthatch, who is facing the wrong direction to see things coming toward him from above. He bends his neck until his bill points straight out from the trunk and cocks his head to look back up. But the creeper is light footed—too careful a climber to create a rain of debris.

When the nuthatch reaches the base of this tree, he flies to the crown of the next. There he pauses to sing—his call sounding like notes blown on a toy soldier's tiny tin bugle. But the forest is still; no other nuthatch returns his call, so he begins his next descent.

The creeper continues less obtrusively to the treetops. From there she plunges through the strata of limbs to the base of another great trunk to continue her perpetually uphill trek. The little bird will walk almost a mile straight up before the morning is over.

The two feathered hunters move off through the forest, and quiet returns to the trees. After a while the first slanting rays of sunlight find the slim, raised flake of bark decorated with the silver crescent. Responding to the warmth, the dark form begins to expand, antennae slowly erecting from their stowed position. Paired forewings rise from between the folded hind wings, unveiling wing margins deeply indented with irregular notches. This jagged edge, which gives butterflies in this genus the name anglewing, shows none of the smooth profile that birds associate with the wings of butterflies and moths.

Anglewings are found in all the world's conifer forests. Each species bears a distinctive silver chevron on the lower hind wing. In some, the imprinted emblem extends to resemble a fishhook, in others a letter or a punctuation mark. It can be broken, suggesting a semicolon or a question mark. In this one, the white form resembles a comma.

The distinctive silver shape stands out against the simulated bark pattern of the underwings—it is not meant to be overlooked. In a bird's-eye-view only an inch away, the bright metallic form is readily perceived and recognized immediately, but for what it is not. When the nuthatch and the creeper focus on the crescent's sharp outlines, the broader, dull, irregular profile of the surrounding butterfly blurs into the background. The bright mark is misidentified and then quickly passed over—rendering the butterfly itself invisible.

As the morning sun strengthens, the mist burns away. To better absorb the heat, the angular wings snap open. Boldly patterned orange and black upper sides glow in the light. In a few moments, the insect will be warm enough to take to the open air. Its flying colors will be visible far and wide, advertising its presence to others of its species—potential mates, and competitors for those mates. Should a nuthatch or creeper lunge at its perched form, the butterfly will evade the attack with a flick of its wings. As it dances into the air, the butterfly's agile, erratic flight will leave the birds with no thoughts of further pursuit. The anglewing will be safer in plain sight than it was hidden on the bark.

Science Notes

Dark, cryptic coloration is found on the upper forewings of the moths and on the under hind wings of the butterflies, respectively—the surfaces exposed at rest by the two families. The anglewings disguise their outlines with an irregular, angular wing margin. They also display a small, bright focal point on the lower hind wing, which distracts predators from the perception of their outlines. The white-breasted nuthatch (Pravosudov and Grubb, 1993) and the brown creeper (Hejl et al., 2002) exemplify the division of one habitat into different niches through the pursuit of distinct foraging strategies. This division is in keeping with Gause's principle, which states that separate species will not compete directly with each other, and that if two species were to try to occupy the same niche, eventually only one would prevail.

References

Hejl, S. J., et al. 2002. Brown creeper (*Certhia americana*). In A. Poole and F. Gill, eds., *The birds of North America*, no. 699. Philadelphia: Academy of Natural Sciences; Washington, D.C.: American Ornithologists' Union.

Pravosudov, V. V., and T. C. Grubb. 1993. White-breasted nuthatch (*Sitta carolinensis*). In A. Poole and F. Gill, eds., *The birds of North America*, no. 54. Philadelphia: Academy of Natural Sciences; Washington, D.C.: American Ornithologists' Union.

Puppeteers

Posed on the rock beside a dry wash, the fly is dead still, impassive, oblivious to its surroundings. Its wings are held at an exaggerated angle as though poised for flight, but even though the morning has warmed, the insect has not moved for long minutes. It is no longer a free being—its actions have been redirected to purposes not its own. Its body is now governed by a miasma spreading within.

A second fly lands nearby. Flies are gregarious creatures, but before the newcomer can make contact with the first, the scene falls under the shadow of a more commanding presence—a hunting jay. The second fly escapes, but the first does not move. The bird ignores the potential meal—she recognizes the unusual pose and the dark flower of spores spreading away across the surface beneath it. She knows that the motionless fly is now unpalatable due to the infection that has killed it—this insect is a victim in the constant contest between the flies and their parasites.

Earlier in the season, when the watercourse ran spring fed and the sun was at its highest, the story would have been different. The fly could have resisted the ruinous fungal spores it had wandered across—it could have shed them from its feet. The air would have been too dry for the spores to survive on the ground long enough for a second chance encounter. Those few spores that did manage to infect a fly would have been slowed by the insect's immune system, which functions best when the sun of high summer heats its body. The insect could have mounted a behavioral fever in midsummer, killing the contagion simply by exposing its black back to continuous sunlight—raising its body temperature high enough to nearly kill itself.

All summer long the flies manage to escape the pathogen. They spend hours rubbing their legs together in hand-washing motions, shedding the sticky spores they chance to pick up when they walk across infested ground. Upon close examination, those fiddling legs appear as fearsome as any spike-studded armor—they are sheathed in saw-blade rows of spines and bristles.

The flies constantly curry themselves, and the spores they comb from their bodies accumulate on their leg bristles. As the legs rub together, the spores are transferred back and forth between them. In the process the spores pick up dust and lose their stickiness, finally falling inert to the earth. Just as a cricket sings by rubbing files against scrapers, so the rhythmic cadence of the flies' regular rows of leg bristles rasping back and forth produces a microscopic trilling chirp, audible only to flies.

The theft of a host's free will is a strategy that allows certain parasites to improve their own success. They cannot kill their host outright, as does a virulent pathogen, because that would break the infection chain. These parasites prosper by getting themselves transferred into their next host before their current host dies. To accomplish this, they manipulate the host's behavior, so that the host works for them to ensure the transfer.

This kind of behavioral parasitism is seen not only in flies, but also in ants, pill bugs, caterpillars, and many more—each victim driven to put itself at risk for the sake of its microbial tormentor. Many behavioral parasites pursue an alternate-host strategy. To complete their life cycle, they must pass between very different animals.

One microbial parasite of ants must find its way into a grazing mammal before it can continue its development. This parasite will not kill either host outright, but it could sacrifice the insect if it could do so at the very instant the infection moves into its next host. The microscopic parasite accomplishes that—by altering the ant's behavior.

While uninfected ants forage in the morning, the infected ones cease their work and climb to the tops of the highest stems of grass. There they clamp their mandibles down on the blade tip and then hold that posture throughout midday. Once the afternoon passes, their internal master allows them to release their bite and return to foraging.

The infected ants repeat this behavior every day—until they have been eaten. They are unintentionally consumed by grazing herbivores that ingest the ants and the parasitic microbes within them, and in so doing, become infected themselves.

In an analogous scenario, a microscopic spiny-headed worm lives a parasitic life cycle that alternates between pill bugs and birds. To accomplish the transfer between them, the parasite impels its arthropod host to adopt suicidal behavior. The infested pill bugs abandon the cover of the leaf litter in which they would normally hide all day, and then they go wandering about in plain sight. If they survive their day's escapade, they are released back to feeding at night as usual, but morning finds them exposed again. Enough of the wanderers are seized by foraging jays to ensure that the parasite they carry is ingested, thereby transferring the infection to the next victim in the succession of hosts.

A class of viruses that infects the colonial caterpillars of hardwood forests takes control of its hosts, causing them to cease feeding, leave the cover of the leaves, and go wandering through the branches in only one direction—up. Those larvae that survive predation by the birds are killed by the virus when they have climbed as far as they can. The dead caterpillars are left hanging by their middle legs in an inverted V. From this position the virus is assured of maximal dispersal of its infective viral progeny, shed from the very tops of the trees over the broadest possible swath of the forest below.

A host animal's reactions and behavior are controlled by its nervous system—an architecture of millions of neurons radiating from its brain. How could a microscopic parasite commandeer such a network—orders of magnitude larger in size and complexity than it is? These tiny pathogens manage no more than their basic reproductive multiplication, yet they have developed the ability to countermand the primal instincts for survival of their much more complex hosts.

They are puppeteers. They afflict all classes of animals. Their strategies are tailored to circumvent the wide range of immune responses and behavioral defenses their hosts may possess. Host and parasite dance with each other over the eons, each adjusting its numbers and tactics to maximize its survival in spite of the other. They strike a balance—the hosts continually refine their behaviors to avoid as many parasites as they can; the parasites modify their multiplication to nondebilitating levels. Host-parasite interactions are a constant feature of life on Earth—predominant in shaping the nature of the living world.

The summer winds down, and after many successful generations, the flies finally run out of heat. The cool, damp conditions favorable to the fungus return with the fall, and the parasite once again prevails. As the days shorten and the parasite becomes established in its host, it takes over the fly's behavior.

Late in the afternoon, while the other flies seek shelter, the infected insect climbs onto an exposed surface. It stops one last time, in plain sight. Gradually, the parasite extends its host's wings to the fully raised position. The fly's last act before sunset is to reach down and kiss the ground at its feet. In this stance, head down, tail up, wings raised exposing the abdomen in mating posture, the microbe locks the insect in position.

In the humidity of the night, the fungus sporulates, covering the moribund host with innumerable sticky spores, shooting still more spores across the area in a radial pattern centered on the hapless insect statuette.

Morning comes. Other flies slowly become active in the chill autumn air, searching for mates, piling on one another. Flying past, a male fly happens to spy the form of a second insect posed in the receptive posture. He alights directly upon it—only briefly—and the cycle continues.

Science Notes

Some microbial parasites improve their own chances for success by manipulating the behavior of their larger host animals (Moore, 1984, 2002). Their behavioral manipulations are seen in a range of hosts that extends from the smallest arthropods to humans. Microbes that alter the behavior of insect hosts (Ray et al., 2006) induce a variety of pathological behaviors, including "summiting" and the adhesion of moribund hosts to the substrate. The behavioral fever and grooming activities are antiparasite defensive reactions of insects (Ray et al., 2006). Fungi in the parasitic genus *Entomophthora* include many that are parasitic on flies (Werren, 1994), as well as on a broad range of other insects. Male flies are attracted more to *Entomophthora*-infected females than to infected males (Zurek et al., 2002).

References

Moore, J. 1984. Parasites that change the behavior of their host. *Scientific American* 250(5): 108–15.

———. 2002. *Parasites and the behavior of animals.* New York: Oxford University Press.

Ray, A. E., et al. 2006. Bizarre interactions and endgames: Entomopathogenic fungi and their arthropod hosts. *Annual Review of Entomology* 51:331–57.

Werren, J. 1994. Genetic invasion of the insect body snatchers. *Natural History* 6:36–38.

Zurek, L., et al. 2002. Effect of entomopathogenic fungus, *Entomophthora muscae* (Zygomycetes: Entomophthoraceae) on sex pheromone and other cuticular hydrocarbons of the house fly, *Musca domestica. Journal of Invertebrate Pathology* 80:171–76.

Part 5. Earth and Stars

The eye of the beholder. In the last instant of its journey, the image of starlight is transformed by its passage through the atmosphere, then through the eye itself.

The Light Fantastic

We live beneath a special sky—a compromise between extremes. Our atmosphere is unique in the solar system, neither an oppressive overcast (like that of Venus or Titan) nor a tenuous near vacuum (like the deep space close upon the surfaces of Mars or Europa). The layer of air that covers our Earth is substantial enough to shield us from the lethal winds of the cosmos, yet it is not opaque—it is transparent. So while our bodies are protected by the blanket of air that covers us, our mind's eye is free to roam among the lights of the night—to rise up and then look back down upon Earth from the perspective of its starry context.

Looking up, we see the lights of the stars through the sea of swirling air currents miles deep that stirs the air above us. The farther from the zenith we look, the deeper the atmosphere is—forty times more air intervenes when our sight wanders just above the horizon than when we glance straight up.

The currents in the ocean of air constantly mingle pockets of different wind speeds and temperatures. Buoyant updrafts fly past chill zephyrs spilling from the stratosphere; eddies spin where zones of wind shear moan against one another. When the light from the stars crosses boundaries between these different regions of air, its path is bent.

Sirius, the Dog Star, stands out among the night's lights as a focal point for our sense of celestial wonder—the brightest star in the night sky. Its light is refracted and reflected from the boundaries between the different pockets of air just as it would be by the solid faces of a prism, though not as strongly. Each of these air pockets acts as a miniscule lens. Their effects are amplified by their vast numbers—even more so when a star is low in the sky where the intervening air is thicker. Fifty miles of atmosphere surges between our eye and the image of Sirius rising just above the horizon.

The fleeting lenses in the air flicker in and out of existence, animating the light of the stars—causing them to sparkle. The effect is greatest on cold, crisp evenings in the hours when the air still roils with the energy of the day as the stars are first appearing. Early in February, Sirius is climbing just above the murk of the southeastern horizon and is already at full brightness when darkness asserts itself. Sirius's shallow rising angle keeps it in the low region of greatest atmospheric effect longer than any other of the bright stars, which rise more vertically toward their zeniths.

Seen through a small telescope at dusk on the winter cross-quarter day (halfway between the solstice and the vernal equinox, about February 5), Sirius appears to swim in and out of focus. Its light dances like the image of a sunlit coin shining up from the bottom of a stream, distorted by the ripples on the surface.

Seen on that day with the naked eye, Sirius the evening star scintillates—its gleam in constant motion. Hovering in the southeast, where the backdrop darkens early under the arc of the rising shadow of the Earth, the Dog Star comes alive, winking to catch our attention. Though the air is never more transparent than through storm-washed late-winter skies, the light of this restless star sometimes disappears, switched off for a split second—then it comes back and sparkles even brighter than its true brilliance.

Suddenly the star flashes electric blue for an instant; a moment later, it flashes with a pulse of ruby red. Our atmosphere is dividing the starlight, parsing the distant beam into darts of color. These are not stellar hues—not the pale yellow of Capella, not the Martian orange cast of Betelgeuse. No star is green; no visible star is bright red. These are bold prismatic colors—the same pure shades we see refracted from dewdrops when the morning sunlight hits them just right—but with an emphasis on the blue. Sirius is a blue star, actually brightest in the ultraviolet wavelengths our eyes cannot see. So its sapphire scintillae are especially prominent—a glittering counterpoint to the deepening gray of a late-winter evening.

Those rays of starlight have been traveling at the speed of light for the last eight years, across trillions of miles of space. Yet the sparkle in their luster is created in the last millisecond of their journey to us, brought to life by their passage across the few dozen miles of air we live beneath.

And in the last inch of its flight, a star's light is embellished for us even further—the ray points are added around its circumference during its passage through the lens of our eyes.

Two viewers standing side by side each see a unique set of ray points around Sirius, since the fine structures of everyone's eyes are unique. Each viewer also sees a unique sequence of colored darts. The refraction of starlight is a consequence of the bending of the beam by less than millimeters, resulting in a different light show for each viewer (and for each eye of each viewer).

The spectacle of Sirius's light show lasts through the twilight, until the star has risen far from the horizon. But when it shines from on high later in the night, or when it first appears higher later in the spring, or in the wee hours before a fall dawn when the air has settled down, Sirius will have lost its sparkle and appears blank and remote, rinsed of color—looking like the dimmer cool white stars above it.

Sirius is also featured on the opposite side of the seasonal wheel—in the heart of summer, when its passage across the sky is invisible. The dog days come to pass when the Dog Star makes its closest pairing with the sun. On July 5 Sirius crosses the meridian—the imaginary vertical line rising from due south—at noon, the same time the sun does. By then Sirius has not been visible for more than a month. At the end of May it draws so close to the sun that it can no longer be seen, even just after sunset—it sets before the light fades enough for it to emerge into view. Like every other star along the celestial equator, each in its season, Sirius spends part of its year living only in our mind's eye, having disappeared from the sky into the blaze of the sun.

In August the summer cross-quarter day arrives—halfway between the solstice and the autumnal equinox. By then the embrace of a clear, chilly winter evening is a memory as distant as a daydream. But early the following morning we can set our eyes on the beacon of the Dog Star once again. It will have drawn far enough west of the sun to emerge through the low fires of dawn for a minute or two before being swallowed in the building glare. With this reappearance, the dog days of summer come to an end and the season begins to turn.

Most stars take longer to reappear, but the light of Sirius is strong enough to penetrate a twilight that overpowers the others. Its starlight strengthens as the summer wanes. Rising four minutes earlier each night, it appears among the low hills brighter every morning against a continually later dawn. Soon it is well up in the southeast, burning at full strength by the time the horizon's silhouette materializes—a sure sign that fall is just around the corner.

The spectacle of Sirius peeking through the flares of daybreak or sparkling fiercely on the evening of the winter cross-quarter day stirs our sense of celestial awe. Many other wondrous stars on the vast plains above recede to infinity before our eyes each night—each one of them as phenomenal as Sirius, or as our own sun. But most are too far away to see—we can as yet only imagine most of them.

The star patterns seen from a planet orbiting Sirius would look much the same as those that appear above our own evenings, with the exception of one extra star in that faraway night sky—a yellow beacon not found among our constellations. This star would be seen from a Sirian solar system in color contrast to a blue star familiar to us—Vega, which would lie next to it. That yellow star—our own sun, seen in the context of the deep background stars against a deeper black sky—may appear as beautiful to the Sirians as the bluish image of their sun appears to us.

And there is another, even more beautiful point of light in the night sky closer to us, but also beyond our sight—living for now only in our mind's eye—invisible both from Earth and from Sirius. This is a variable star, one that changes through many orders of magnitude over its cycle, with a maximum brightness many times greater than that of Venus, our brightest evening star. Venus would provide a particularly good vantage point from which to see this gem, if the atmosphere there were less opaque. When gaps open in the Venutian clouds, this beautiful point of light stands forth waxing and waning across the seasons, its color changing as well as its brightness. It evolves from night to night, shifting unpredictably between shades of ivory white and deep luminous blue.

Here on Earth we are deprived of the chance to view this special evening star not because we are too far away but because we are too close. We are standing on it. The changing spectacle that Earth presents to a distant vantage point can only be imagined from here. We can see our starry context every night, but we cannot rise and look back down upon our planet against that context. Its blue beauty laced with passing white clouds floating before the Milky Way exists for now only in our imagination.

The spectacle of Earth as it would appear set against the black backdrop of its native environment would offer us a welcome change in vantage point. One day, such a big picture will be easily accessible for all to contemplate. Until then, we might pause when we look up at the stars twinkling in the evening sky and just imagine that sight—and the alternative perspective it would offer on our earthly concerns.

Science Notes

Small, invisible pockets of air at different temperatures and wind velocities impart slight reflections and refractions to the paths of light beams passing through them. Refracted light is broken into its spectral colors, which add the flashes of green or turquoise to twinkling starlight. These are prismatic colors, not star colors. (Carbon stars are deep red, but none are close enough for us to see their color without telescopic enhancement.) Turbulence aloft builds as the day warms; then it subsides as the heat radiates away again into space at night. Clear rivers of air can flow stronger in the cooler seasons. The twinkle strength of the atmosphere changes from night to night. Sea breezes or the presence of the jet stream overhead can animate starlight. The air gets warmest during the dog days of summer, but that air can be heavy and calm, having little animating effect on starlight.

The "dog days" here are defined in terms of proximity of Sirius to the sun. An almanac definition might specify "the forty days beginning July 2 and ending August 10." The term is a folk saying associated with the doldrums of summer; it originated in the time of the pharaohs, when the dog days began in midsummer—after Sirius's heliacal (sun-coincident) rising day—with its first appearance in the morning sky. Seasonal stellar positions have changed since them, reflecting their alteration with the precession of the Earth's axis. In the present epoch, Sirius can last be seen at sunset around May 30; it transits directly behind the sun on July 5, and it can first be seen at sunrise around August 1 (depending in part on the persistence and acuity of the observer).

Twilight is divided into periods when the whole sky is said to have a certain brightness, or magnitude. The sky in the first period, civil twilight, has a magnitude that precludes visibility of all but the brightest stars. Sirius, with a magnitude of −1.4, should be visible through a twilight that would hide other stars. (Venus, magnitude −4, is visible through daylight.) Stellar brightness is measured on a magnitude scale: the larger the number, the lower the brightness. Each successive increment is 2.514 times dimmer than the last.

Stars on the celestial equator spend 10 percent of their time hidden behind daylight, when the annual motion of the sun has come to match their longitude. The rotation of the starry dome advances relative to clock time by about one degree (four minutes) per day. Northern stars, such as Vega, are lost in daylight for less time. At midnorthern latitudes, Vega appears through the morning twilight on the same day it disappears in the sunset—it is visible every night of the year. Stars positioned south of the celestial equator, such as Fomalhaut and Sirius, are lost for longer—hidden for some time behind the Earth as well as the sun. The day of the summer solstice—when the sun reaches its greatest altitude—is the longest of the year. The days are approximately that long (+/− 5 minutes) through July 5.

Even though air is transparent, it becomes translucent and then opaque to transmitted light as its thickness increases with a viewing angle approaching the horizontal. Limiting visual intensity decreases as the (log of the) mass of the intervening atmosphere increases. Stellar magnitude diminishes in proportion to the secant of the angle of departure from the vertical, as derived from Laplace's theorem of the extinction of starlight. That trigonometric approximation holds until the viewing angle closely approaches the horizontal, where the curvature of the Earth and low-angle refractions alter the calculations. Thus, the stars never actually set beneath a flat horizon (oceanic, desert) but disappear above it—the dimmest stars fading the highest (five degrees above). The gloom of the nighttime oceanic horizon contrasts with the starry dome on high not because of "sea haze" but because of the extinction of light by the atmosphere.

Sirius is relatively close to us—eight light years away in the direction of Canis Major, the Big Dog. Another bright winter star is Procyon—eleven light years away in Canis Minor, the Little Dog. Their lights are transformed into sparkling gems during the last quarter of a millisecond of their years-long journeys, while they pass through our atmosphere. The circumferential spikes on the star are added as the light passes through our corneas; that passage takes about a picosecond (one millionth of a millionth of a second; 10^{-12} second).

Sirius actually shines brighter south of the equator, where it rises to the zenith and its light is obscured by less intervening air. But it sparkles less and throws fewer colored darts for southern hemisphere viewers. Its colorful spectacle is best beheld by those in the North for a few days in early February, when the star lingers low in the last of the southeastern dusk—an evening beacon rising through the tides of our ocean of air.

Earthlike planets orbiting Sirius in the habitable zone inward from the orbit of Sirius's companion star (the white dwarf Sirius B) have not yet been discovered. A seasoned star watcher come to Sirius from Earth would detect subtle differences in the starry sky above a planet orbiting Sirius but would still recognize all the familiar constellations. Our sun would appear there as a yellow star twenty-two degrees from the blue star Vega. The sun would be less than half as bright (magnitude 1.8) as Vega would appear (mag. 0.6; dimmer than the mag. 0.0 at which we now see it). Using our current technologies, the Sirians looking back at our sun would detect no signs of habitable planets associated with it.

Gold

The gold nugget rests in your hand, still wet. Its lustrous surface reflects a world unique among the stars—your world, a place that has the power to produce this glittering gem from base rock. That alchemy requires enormous pressures and temperatures, millions of years to carry out, and a special setting as well—as far as we know, it could only happen here.

Your nugget was once gold dust scattered at infinitesimally sparse concentrations across the void of outer space. Gold constituted less than 0.0000005 percent of the matter that consolidated from star dust into the primordial Earth—a negligible trace constituent of our growing sphere. But Earth was destined to be a water world, and that would bring sweeping change.

Oceans now rest uneasily on the thin rock firewall covering Earth's core, where the incandescent heat of its creation is retained. The interplay of the fires below and waters within the crust has energized a dynamic geology that drives the tectonic motions of Earth's surface. That motion has created folded ridges tens of thousands of feet high, still showing the sedimentary strata on which they were built. Those strata were assembled at the bottom of the sea, and as they grew, they came to be shot through with veins of crystalline ore and sprinkled here and there with pieces of solid gold.

The fires beneath the mountains show through the Earth's crust where it is thinnest—at the bottoms of canyons deep beneath the seas. There, newly created seafloor spreads away from its origins—a broad conveyor belt following a path that will lead to its eventual destruction. Seafloor is destroyed at the end of a multimillion-year migration, when it slides into canyons steeper still, meeting the fires of its birth again deep beneath the beaches of the continental margins.

At those regions where the crust is created and destroyed—beneath the centers and the edges of the oceans—groundwater exists in a form incomprehensible to those who know only the surface of the land. Deep underground, water has been forced into contact with the Earth's fiery magma, where it grows so hot and pressurized that it dissolves quartz. It also dissolves gold.

Gold is not soluble in water as we know it. Where we live, a gold ring will stay in intimate contact with the skin moisture of a finger for a lifetime and never tarnish. But at depth, gold readily dissolves and leaches from the rocks. It is scoured one atom at a time from cubic miles of the Earth's crust by the superheated solution percolating through the cracks. This is the first step in its concentration.

The plates of the Earth's crust carry their mineral-laden groundwater with them as they move away from the volcanic boilers where seafloor is created. The temperature of the hot groundwater declines as the volcanic fires are left behind. And as the solution cools, it undergoes a series of transformations.

Each of the minerals dissolved in the buried brine cools to its own critical temperatures, and then each materializes in solid form. They appear as crystalline deposits on the walls of flooded fissures. These crystals grow pure—self-selecting their own atoms one at a time from the solution. Their formation concentrates dissolved manganese into pink seams in the rocks. As the temperature continues to decline, brown bands of iron oxide are deposited. The process continues, and veins of solid gold grow on surfaces previously varnished in translucent quartz. Over the millennia, all the dissolved atoms in cubic miles of groundwater are purified from each other and concentrated into clefts in the rock. This creates lodes of mineral deposits that remain in place when the water finally runs away.

Some continental margins are particularly well suited to the accumulation of gold. Among them are the mountain ranges behind the shores of the eastern Pacific—the Andes, the Sierra Nevada—terrains shaped both by rocks derived from seafloor sediment and by active volcanism. The sedimentary strata there are the remnants of generations of mountain ranges long since reduced to sand, washed down on each other and compacted into rock, then uplifted tens of thousands of feet above the level of their creation. Harder mineral-bearing blocks shaped by volcanic action have been sutured to those sedimentary strata during their tectonic uplift.

The thundering energy of the water planet we live on is readily apparent to anyone hiking past the cataracts that sculpt these grand cordilleras. Erosive removal of the softer limestones and shales gradually uncovers the underlying mineral-bearing veins.

Once exposed, most minerals are weathered away—oxidized by the air and dissolved in the rain—but not the gold. Gold initially deposited in the rocks under the seafloor is born in its metallic form. As such, it does not react with other elements—as does copper, for example, which combines with sulfur to form green ribbons of copper sulfate, or mercury, which occurs as cinnabar veins of mercuric sulfide. As a nonreactive element, me-

A lucky day. A rounded gold nugget finds rare repose in plain sight in a streambed. Nuggets remain buried throughout most of their existence.

tallic gold is resistant to the chemical weathering that destroys deposits of other elements.

Erosion finally removes the veins of gold-gilded quartz from steep bedrock walls and drops them to shatter against the rocks below. The golden fragments do not lose their integrity when removed from their encasement. Their strength holds them together as they tumble downslope with the rest of the broken pieces of the mountain. Carried along through the low points by running water, they survive intact while softer debris is reduced to clay. This process ultimately collects all the gold nuggets from an entire cliff face into a few deep mountain pools.

The crests of the mountain range reach into the stratosphere, but even so, they are mere shadows of what might have been. Mountain ramparts would reach thousands of feet higher were it not for the forces of destruction that oppose their growth. At a rate that is too subtle to be appreciated by a walker pausing on an alpine trailside, the highest ridges may lose a thousand feet of height over ten million years of erosion—which is just a fraction of the age of the range.

But while the mountains crumble, the gold remains. Though the rocks in which it was uplifted have been removed from around it, the gold stays near its original position. It rests in pieces like shards of a broken bottle still scattered on the spot where it shattered, long after its less enduring contents have trickled away.

All the gold from a long-lost mountain wall will have accumulated in a latticework of low points—the traces of the succession of streams that crossed the rocky face during its erosive destruction. The metal will have sunk as those streams ate away the rock; gold becomes enriched in deposits that form along the meanders of fossil flow lines—collapsed into a flat shadow of their once three-dimensional network.

The gold stays in place because of its unusually high density. When the force of the streams strengthens at the height of the spring runoff, the gravel and the lighter minerals parade off with the current, rolling and bouncing downstream. Flashy gold-colored flakes of iron pyrite tumble past, but the true gold lags behind. It descends in elevation only when the rock below it is removed.

Gold eventually finds its way into the deepest depressions worn in the streambed. When the force of the hundred-year flood reaches into those holes with enough power to stir the rocks around, the gold is sorted to the bottom of the pile. It shares this mineral niche with a few other dense lodestones and gemstones that have been collected by the same process.

As the millennia cascade past and the mountains decline, the gold descends to calmer waters. The nuggets have by then had their rocky crusts removed and rough edges rounded by the glancing impacts of thousands of passing stream stones. Gold may come to reside on gravel bars in languid stretches of foothills flat-water. Salmon spawn among the nuggets; passing moose and ground sloths pay them no heed as they trample across the shallows.

The nuggets collect close to the bedrock and are covered ever deeper by the less dense overburden. The entire layer is finally compressed into sedimentary rock that may ultimately be lifted back up again. Then the cycle continues—buried placer deposits of heavy metals and gemstones will be unearthed again when new streams arise in the new mountains.

The mountains produce gold nuggets continuously, though rarely. That gold slides downslope across geologic time, spending most of its days resting in places the sun never reaches. Those moments in time when a stream-borne nugget chances to be exposed to the open sky—pausing in its journey from one deep resting place to the next—are rare indeed. Rarer still is that moment that brings you to cross the mountain stream at that very spot where a nugget happens to lie shining on the gravel.

Science Notes

Earth is a rare world—the only one we know of that sustains liquid water on its surface (though we would like to know of other such and are actively searching for them) (Raymond et al., 2006). Liquid water permeating our planetary crust affects the ductility of the rocks, a condition conducive to moveable plates (Mackwell et al., 1998). There is scant evidence of plate tectonic activity elsewhere in our solar system, and, by extension, the rest of the universe. Mars and Venus appear to be coated in single-piece shells (one-plate planets) (Loddoch et al., 2006); shield volcanoes on Mars do not form chains like those they form in the Hawaiian archipelago—indicating no Martian motion of the plate they erupt through. Mercury also appears superficially static, as does our moon. There is evidence of past movements of the solid crusts of some of the moons of the outer planets, but they too appear now to be generally frozen in place. It may be that the ductility of a crust permeated with water is essential for a dynamic surface—for example, the anhydrous, brittle crust of Venus, Earth's twin, may resist tectonic deformation (Mackwell et al., 1998)—which would put our water world in a tectonic category by itself.

The atoms of gold that make up gold nuggets are billions of years old. It has taken but a tiny fraction of that time, some hundreds of millions of years, to forge them into metal and move them to the surface of this planet. Gold is the only element likely to be encountered among the rocks in metallic form. Its atoms were initially diluted evenly throughout our sphere—igneous rocks are about five ten-millionths of a percent gold (Hill, 2006), about one atom for every million atoms of silicon. The dynamic surface of Earth eventually concentrates gold into primary hard-rock lodes and secondary sedimentary deposits. Such sedimentary deposits, called "placer deposits," form when heavy minerals (gold, platinum, chromite, metallic copper, magnetite or lodestone, ilmenite, zircon, ruby, diamond, sapphire) separate from less dense materials as they are carried downstream after weathering from their initial hard-rock deposition sites. (The same process of density separation allows the collection of gold by panning.) Gold is among the rarest of these minerals—all the gold ever mined would fit into a cube twenty-five meters on a side.

The longest mountain range in the world is unexplored. This is the continuous midoceanic ridge system. Its peaks look down into deep canyons where seafloor is produced from rising magma. Seafloor spreads away, carrying these mountains across the globe. This mobile terrain is eventually destroyed, sliding into the deepest canyons in the world—also largely unexplored—the coastal trenches that mark zones of crustal subduction.

"Critical point phase transition" is a condition of temperature and pressure above which a material dissolves in water but below which it becomes insoluble and separates out into crystals. "Differential crystallization" describes the process by which different elements or compounds are deposited as solids at different times as a body of groundwater cools. The elements have distinct phase transition points, thus differentially purifying themselves away from the mixture of dissolved minerals. Lodestone is magnetite, an iron-bearing ore. Fool's gold is a shiny yellow form of iron pyrite (iron sulfide).

Erosion is a constant force, continually bringing down the mountains. If we say the rate is one thousand feet of height destroyed per ten million years, this gives us a sense of the contest between elevation and erosion for a mountain range ten thousand feet high. The pace of erosion is not constant but depends on many factors; it is faster in steeper, taller growing mountains than in older, lower ranges. Placer mining was the most massive environmental force in recent geological history to shape the Sierra Nevada. Destructive gold mining continues today in Central and South America. All the gold will never be extracted—more will always remain, though always less than before.

References

Hedenquist, J. W., and J. B. Lowenstern. 1994. The role of magma in the formation of hydrothermal ore deposits. *Nature* 370:519–27.

Hill, M. 2006. *Geology of the Sierra Nevada*. Berkeley and Los Angeles: University of California Press.

Loddoch, A., et al. 2006. Temporal variations in the convective style of planetary mantels. *Earth and Planetary Science Letters* 251:79–89.

Mackwell, S. J., et al. 1998. High temperature deformation of diabase with applications to tectonics on Venus. *Journal of Geophysical Research* 103:975–84.

Raymond, S. N., et al. 2006. Predicting planets in known extrasolar planetary systems. Pt. 3: Forming terrestrial planets. *Astrophysical Journal* 644:1223–31.

Simmons, S. F., and K. L. Brown. 2006. Gold in magmatic hydrothermal solutions and the rapid formation of a giant ore deposit. *Science* 314:288–91.

Incandescent Falls

Incandescent Falls cascades from the bluffs on the south shore of Hawaii Isle, tumbling directly into the waters of the tropical Pacific. It is a constant presence on this midocean archipelago, though it alters its flow and position from time to time, conforming to the ever-changing islandscape around it.

On those rare nights when it flows at maximal strength, the cataract is mesmerizing to behold. The glowing torrent issues from the mouth of a cavern, and the arc of its momentum carries it out over rocks burnished in its reflected light. It continues across the narrow beach, slipping beneath the shoreline waves, the color fading from its base as it disappears below curtains of mist.

Despite its massive volume, the luminous falls hisses only softly as it cascades off the island's edge—its earthshaking potential muted. It does not sing with the sparkling voices of a stream tripping over rocks because it does not carry water. Incandescent Falls carries magma—molten basalt conveyed directly from the Earth's interior—lava that flows with the speed of a waterfall but is much too dense to splash and burble like a free-falling forest cataract.

This firefall is flame yellow throughout. On its surface it floats rock that would itself glow if set on the sand but that appears black against the even hotter river that carries it. Should the pulsing current throw a few of its golden drops up onto the face of the bluff, they will cool to a dull red as they solidify in place, building out the seawall.

Standing to the side, you feel the heat of the Earth's core on your cheeks and the palms of your outstretched hands. You hear the surf hiss and steam where it splashes against the cascade; flotillas of little pumice rafts are spawned from the quenching reaction, each raft cradling in its center a small glowing ember. Sputtering and sizzling like water dripped into a hot frying pan, they bob by on the waves as they set out across the sea for Polynesia.

This uncommon cataract flows down the bluff at the edge of a landscape as surreal as the radiant river itself. For half a mile back from the bluff, the

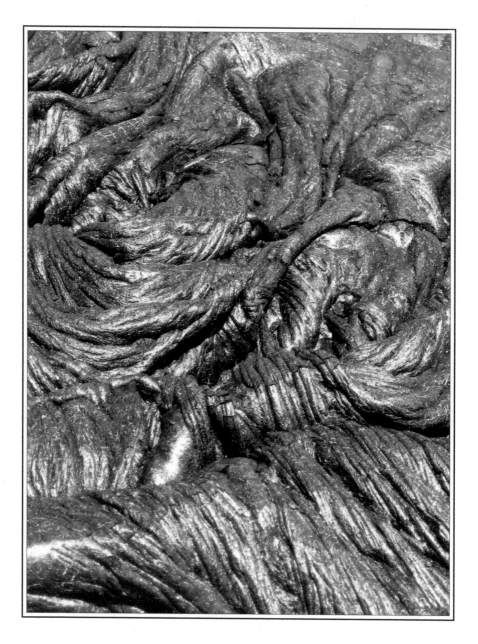

Frozen falls. Ripples in a pahoehoe lava flow have turned to stone.

terrain is flat black lava flow. This ground breathes and moves beneath a smooth, dark skin throbbing just below its surface with a system of arteries that carry the hot volcanic blood of a living landscape.

Those pulsing arteries are lava tubes. They form around streams of lava that snake to the ocean from craters up the slope behind the expanse of seaside tableland. The ground below these lava flows is heated to 1,600 degrees, but the exposed top lava surfaces lose heat to the breeze that washes in from the ocean. The upper crust darkens and hardens, eventually solidifying while the molten rock continues flowing underneath it.

Where the streams of lava level out, approaching the shore, the flow slows. The cooling upper layer consolidates and thickens, eventually forming an arch that roofs the length of the incandescent river—containing the flow in a stone conduit. The magma within is insulated from the air, so it stays hot and fluid. It can course unseen through these stone subways for miles, far from its sources on the slopes to the northwest, never stiffening until it meets the sea. Should the red-hot wells of magma above cease flowing, the lava would drain away, leaving empty tunnels in the rock.

A broad span of this rugged terrain precludes easy access to Incandescent Falls. Few actually witness the hypnotic rhythm of its surge as it issues from the mouth of a lava tube to blaze down the face of the bluff against the night. The nearest access road has been permanently buried under kilotons of spreading lava flow. Approach is possible only by a precarious trek on foot across the black plain. The surface of this lavascape alternates in texture, much of it slippery smooth, silvery black billows of ropy lava that some enjoy walking on barefoot. Elsewhere are brittle black cinders that quickly shred footwear, let alone feet. But after the hike across this plain, one may find the falls quiet—they are intermittent and sometimes don't flow at all.

That hike is a dangerous traverse across the back of an ancient, restless, buried beast whose size and strength are not immediately apparent. A glance back from the sea reveals the shoulders of Mauna Loa volcano, the biggest mountain in the world. But its full magnitude is not obvious—most of it is under water. The violence of its construction is betrayed only by the faint wisps of steam drifting from the distant ramparts of Kilauea Caldera on the mountain's flank.

Nevertheless, the lava beds stretching flat from the base of the slope to the sea are unstable—they occasionally sigh and heave. They demand extreme caution of anyone who might venture to cross them.

Should you be night walking on this treacherous terrain by flashlight, you may notice a gust of unusually warm air mixed with the sea breeze—a sensation like heat waves rising from pavement on a hot afternoon. With that cue, you should kneel down and feel the ground to see whether it retains the burning heat absorbed from the sun by day—a heat the rocks should have given up soon after sunset.

If the rocks are still hot—turn back. Ignore that cue and continue on until your shoes begin to melt, and you are making a big mistake. Thrumming beneath your feet, the lava current erodes the upper walls of its tunnel—leaving them thin in places, heating them until they soften. Walking across such a spot as the ground bends under your weight may be the last thing you do. The roofing over lava tubes is prone to collapse, creating skylights that present a direct view of the incandescent river only a few feet below. That span of incredibly hot flow will have swallowed the section of the arch that fell in from above, along with anything that may have been standing on it.

This terrain is subject to collapse in places above the lava tubes, but on occasion, the entire bench above the bluff may fall into the sea without warning. The shoreline plain is built on brittle lava quenched by the surf before it could conform to the gradual slope of the submarine mountainside. Such a bluff extends out upon the sea surface, weighing ever more heavily on its fragile foundations until it fails to support its own weight and succumbs to the pull of gravity.

Should you be out there on the night of such a collapse, you will be treated firsthand to a magnitude ten earthquake. It begins with a growling rumble beneath your shoes, halting your search for a way around a series of fissures parallel to the shoreline that block your path. The tremors increase until you can barely keep your balance, and the fissure just before you releases a tremendous bang and begins to yawn open.

The ground lurches, pitching you off your feet—your flashlight falls and bounces away sideways across the tilting surface into a second crevasse opening behind you. The light disappears, engulfed by a rush of seawater that surges through the chasms widening on both sides. You scramble up the steep surface and see by moonlight that the entire area has been converted into rows of shifting, house-sized blocks of rock awash in a rising tide. Beyond that, you see a tidal wave churning toward you from the open ocean. Walls of water rushing in from both sides clap together just over your head, pounding you away across the rocks into a swirling chaos of bubbles and darkness.

You survive this catastrophe only if you prepared for the risks of your trek by wearing a life vest beneath your windbreaker. It inflates on contact with seawater, ripping the zipper open across your chest and pulling you back toward the air. The surface is a chaos of waves moving in all directions, strobe lit by lightning that sizzles and pops with the sounds of electric arcs. Shock waves traveling through the water compress your chest just before blasts of compressed air thunder over the surface, driven by detonations somewhere off in the darkness.

The active lava tube has collapsed into the sea along with the rest of the shoreline, resulting in a series of huge explosions where hot magma is rapidly mixed with cool water by the steam expanding beneath it. Boulders fly through the sky, some trailing showers of fiery lava sparklers. These missiles raise giant waves where they crash into the water around you, adding to the chop that throws you back and forth like a bobbing cork.

But soon the commotion subsides. The turbulence on the surface resolves back into the gradual swell of the open ocean, and through the dissipating steam you notice the silhouette of Mauna Loa eclipsing the stars to the north. Far down the new shoreline Incandescent Falls flows on, uninterrupted. Its newly placed incarnation, on a new bluff farther back toward the land, spills a trail of sun orange light out toward you across the water. You orient yourself to the sound of the breakers a hundred yards away in that direction and start swimming.

The periodic collapse of the unstable outflow plain sends blocks of volcanic rock rolling down the submarine slope toward the seabed twenty thousand feet below the surface. These blocks form the foundation on which the dark shoreline continually advances out to sea. Incandescent Falls moves its position accordingly, falling back, then building its way forward again as the island grows. The volcanic source of its fiery river ebbs and flows from the upland peaks, sometimes quiescent, other times throwing up fountains of lava. New vents split the flanks of the living mountain, and the magma streams in new directions. The shoreline outfall ceases in one spot, later to begin again farther up or down the black sand beach. Over geologic time, the falls move from one island to the next as the archipelago of volcanic peaks extends itself to the southeast.

As the roots of the volcanoes move away from their older positions, the older vents grow quiet. The Hawaiian archipelago is their legacy. To the northwest of Hawaii, West Maui Mountain—once the easternmost volcano in the chain—now stands silent, extinct and already deeply gorged by tropical rains. Haleakala—the dormant volcanic crest across the saddle to the east on Maui Isle—has not erupted in recent times; the last flow, from a dike emergent on its southwestern flank, reached the shoreline in the 1800s.

Riding on the waters across the channel from Maui Isle, Mauna Kea and Mauna Loa dominate the silhouette of the Big Island. They are already quiescent enough to be crowned with the silver beads of astronomical observatories. Kilauea Crater is the source flow for the current incarnation of Incandescent Falls, on the southern flank of Mauna Loa.

Farther to the southeast, submarine eruptions build Loihi Isle. Already this mountain is the size of a typical terrestrial volcano, yet its peak is still thousands of feet below the surface—it is only a bump on the massive shoulder of Mauna Loa. It is built from a form of lava not found on land—pillow lava, which is quenched continuously from the instant it emerges—shaped by seawater at such pressures that steam generated at the outflow remains a liquid. After growing a few thousand feet taller, after a few more million years, Loihi Isle will break the surface, while the volcanic fires grow dim on the Big Island behind it. Soon thereafter, Incandescent Falls will emerge above the newly forming beach there, spilling its trail of sun orange light again out across the tropical sea.

Science Notes

Lava flowing from Pu'u 'O'o vent on the south side of Kilauea (Moore et al., 1973; Decker et al., 1999) has buried the Chain of Craters road along the southern shoreline of Hawaii Isle. That lava congeals in two forms: either slippery smooth, silvery black billows called pahoehoe lava or, when it flows colder with the surface already solid, a coarse black pumice called 'a'a lava. The Lae'apuki lava bench collapses regularly, most recently in July 2006. During a collapse in 1993, a walker in the area was lost and presumed dead. Active lava tubes (Peterson et al., 1994) that run through regions that collapse into the sea produce steam explosions where magma and water meet in large amounts (Mattox and Mangan, 1997); the explosions throw hot rock a hundred yards back up onto the shore. The steam is darkened by rock pulverized by the explosions; low-altitude lightning can be generated in the cloud from the volcanic dust rising from the water.

References

Decker, R. W., et al., eds. 1999. *Volcanism in Hawaii*. U.S. Geological Survey professional paper 1305.

Mattox, T. N., and M. T. Mangan. 1997. Littoral hydrovolcanic explosions: A case study in lava-seawater interactions at Kilauea Volcano. *Journal of Volcanology and Geophysical Research* 75:1–17.

Moore, R. B., et al. 1973. Flow of lava into the sea, 1969–1971: Kilauea Volcano, Hawaii. *Geological Society of America Bulletin* 84:537–46.

Peterson, D. W., et al. 1994. Development of lava tubes in the light of observations of Mauna Ulm, Kilauea Volcano, Hawaii. *Bulletin of Volcanology* 56:343–60.

Window on the Sky

The flattened crystal facets of a free-falling snowflake form a hexagonal window. That window looks out over a panorama of unparalleled depth. If you could tumble across the sky beside a snowflake, hovering just above its surface and looking through, the vista you would see filling the frame would be breathtaking.

The view swings in continuous motion through 360-degree spins and reversals, flipping and rolling as the flake flies, panning across the cloud-scape canyons inside a thunderhead. One instant would focus on the heights miles above—with blue showing through the cracks in the ascending ramparts—and the next instant would look down into the abyss, darkening far below toward the storm-shaded ground. Bolts of nearby lightning fill the frame with momentary flashes of pure white. Then the panes of ice glow in shafts of sunlight that chase prismatic bright lines around the margins of every facet.

The snowflake's six arms frame a constantly changing picture, while the framework of the snowflake itself also changes constantly, all the arms evolving in unison. A snowflake is a manifestation of the fleeting microclimate in which it is immersed. It is not an impervious diamond, aloof from its surroundings; its structure reflects an intimate relationship with an ever-changing environment of which it is an integral part.

The snowflake continually grows new projections and then retracts them, according to conditions around it that alter with every passing instant. Snowflakes live on a thin edge between extremes of temperature, pressure, and humidity. Water does not exist as a liquid in this coincidence of circumstances but converts directly back and forth between crystalline solid and invisible vapor—forging the snowflake's feathered blades to reflect the changes of the moment.

The snowflake is a single crystal of millions of millions of water molecules, each one fitted into the geometric ice lattice in its turn. These miniscule building blocks maintain their geometric order as their formations grow—until the molecular geometry is manifest in the symmetry of the visible flake. Variations in the chill environment dictate whether growth will be at the cusps of primary arrow points or along rows of secondary projections—on the edges of internal margins or thickening filigrees on terraced surfaces. Rising and falling humidities move the structure either to spread its frosty tips or to reverse the process, pulling in its branches and contracting into a simple hexagon that shrinks until it vanishes.

The dynamic flecks of ice sublime away to nothing, but a moment later the process reverses. Countless minuscule hex crystals flash into being in midair, forming around invisible motes of airborne sea salt or dust powder, bits of diatom skeleton, or smoke particles. The branches of the new flakes bloom into arrays of struts and braces as free-floating water molecules condense directly onto the ice matrix, following a formula for endless variation.

The frozen asterisk is only a millimeter or two wide, small enough that each of its rays experiences the same microclimatic shifts. Conditions vary widely between wind-shear zones, inversion layers, and wisps of cloud reaching into sunlight. But any of these unique patches of air is broader than a snowflake is wide, so the consequences of each applies uniformly over every flake—the embellishments on each arm are repeated exactly on all the others.

Countless gradations of microclimate inhabit a single cumulus cloud. Each snowflake ricochets along a chaotic trajectory from one chill eddy to the next. The thunderhead may span temperatures ranging from zero to fifty below, and pressure gradients between five thousand and fifty thousand feet—all connected by shafts of wind whistling past each other at five thousand feet a minute. Two snowflakes do not fly side by side for long under these conditions. Snowy columns of wind collide head on and shatter into countless diverging currents, scattering flakes in every direction. Despite their vast numbers, no two snow crystals follow the same path, so no two of them ever grow to be exactly alike.

Calmer circumstances anticipate the creation of snowflakes—conditions that foretell the arrival of a storm still hours or days away below the western horizon. The leading edge of such a cold front advances along the top of the stratosphere—through an environment like nowhere below. The air pressure there is very low—breathing would not be possible. Water at the temperature of a human body would quickly boil away to vapor; then the vapor would radiate away its energy to space and chill to ice temperatures.

Water vapor molecules borne on the west wind freeze into microscopic ice crystals in the vanguard of the approaching front. The first of these are simple, unadorned hexes—tiny, far apart, and far above the ground. We

The view through a snow crystal.
No two alike.

cannot see them from below, but we can detect their presence—by the way they bend the light falling past them from above.

One of the first signs of a shift in the weather is the appearance in the clear sky above the morning sun of an upside-down rainbow—a circumzenith arc. This halo is formed from sunrays that fall through the flat faces of hexagonal ice prisms and exit through the sides. The path of the light is bent, and the beam splits into its spectral colors as it bends. As the number of ice crystals grows, the first cirrus clouds materialize. The prismatic colors of the circumzenith arc are most intense where they are carried on the streamers of these ice clouds.

The change in the weather becomes more apparent when the sky grows hazy with light scattered from the enlarging flecks of ice. Then a new spectral arc appears closer to—and encircling—the sun. This aureole is formed from sunlight that penetrates one side of floating crystals, moves across the diameter, and exits from the opposite side—a short trip that bends the light twenty-two degrees from its course.

The midday sky fills with identical ice crystals, all tumbling in freefall. All of them bend and color the light the same way when they chance to rotate into the proper orientation. But we see the rainbow colors of the light they split only from those crystals that happen to lie twenty-two degrees from our line of sight to the sun.

As the afternoon deepens and the clouds build in the west, the color drains from the ring around the sun and pools in two spots on the halo's circumference opposite each other on a horizontal diameter across the circle. These two spots are sun dogs—parhelia. They have captured all the light from the rest of the arc because the ice crystals have become aligned. The crystals have grown heavy enough to settle through the sky, and now all of them fall with their flat faces horizontal. We can see the light bent only by those that fall even with the sun from our perspective.

Each parhelion shines a reddish yellow, the cooler colors of the spectrum having been subtracted from a sky now shaded toward the warm hues of the sunset. The sun dogs may disappear from one side or the other as the gathering clouds eclipse the higher layers of the atmosphere. And as the stratospheric humidity continues to build, the growing crystals finally begin to sprout their spiky arms.

The approaching storm carries a vast reservoir of energy—solar energy—absorbed from a distant ocean and transported in the form of vaporized water. Enough water vapor to grow a one-hundred-thousand-ton cumulus cloud stores a monumental amount of power, which keeps the water molecules flying free. But when the air mass chills as it moves over land, the heat in the motion of the water molecules is given back to the sky—the vapor gives up its energy when each molecule loses mobility by binding to an ice crystal. Heat captured from a clear day over the sea is released into a stormy sky over the land. The liberated energy takes the form of wind.

The winds that carry snowflakes throughout their clouds can grow furious in the largest thunderheads. Energy drawn from the formation of innumerable snowflakes builds gale-force updrafts that drive the cumulus crown tens of thousands of feet up to the top of the stratosphere. There, all the water vapor condenses into flakes in the high deep-freeze, and the swarms of crystals—heavier than the air from which they were born—top out and then begin to sink.

Bright sheets of falling snow descend into the shadows of the cloud tops, trailing mare's tails across the sky as their leading edges sweep into regions of lower wind velocity. The intricate crystals move through their continual changes as they slant through the cloudscapes for miles across country—twirling and jostling until they have fallen close to the ground. Then the implacable resistance of the approaching solid barrier stills the whistling wind. Gravity takes over as the predominant force in the calmer air, and the flakes stop tumbling and finally settle flat, rocking gently side to side.

A landscape shrouded in monochrome silence flattens as it rises to meet the sinking snow. The moving picture in each crystal window slows its spinning, then steadies. Finally the frame freezes, fixed on the sky overhead. The only motion left in the hexagonal window is the distant drift of the clouds far above. The free-flying life leaves the snowflake, and the picture in its softening arms fades to white.

Science Notes

The activity of the weather is driven by phase-state conversions in atmospheric water, which is less than 1 percent of the mass of the atmosphere. The relationship between the molecular interactions of water molecules and the behavior of clouds is known as "cloud physics" (Young 1993).

Wilson "Snowflake" Bentley (1865–1931) originated the idea that no two snowflakes will ever be alike. A snowflake weighs about a microgram and so contains about a million trillion water molecules.

The growth of flat snowflakes competes with the growth of long pencil crystals. Conditions within cumulous clouds favor the growth of an initial tiny hex crystal outward along its margins, producing a broad, flat flake. Conditions in cirrus clouds, favor the growth of the seed hex perpendicular to its equator, so that the hexagonal perimeter does not expand, but the crystal grows lengthwise, producing the long thin pencil crystals that spawn sun haloes on clear days. Parhelia appear at the height of the sun and parallel to it above the horizon. They appear in the evening when hex crystals are aligned against the wind pressure generated by their own settling motion through the sky. They also appear in the morning, their hex crystals aligned by rising air currents.

The growth of crystals in the atmospheres of other planets generates forms as fantastic as our own snowflakes—from minerals other than water. These otherworldly crystals have their own unique symmetries and produce their own particular set of sky haloes. The symmetry of the carbon dioxide seed crystals in the Martian atmosphere is not hexagonal but cube-octahedral. Extension of the rectangular faces of that seed—or, alternatively, the triangular faces, depending on conditions—produces three-dimensional snowflakes above polar glaciers of dry ice one hundred degrees colder than our own arctic.

References

Libbrecht, K. G. 2005. The physics of snow crystals. *Reports on Progress in Physics* 68:855–95.

Young, C. 1993. Diffusion growth of ice crystals. In *Microphysical processes in clouds*, chap. 6. New York: Oxford University Press.

Liaisons to a
Rare Planetary Alignment

The dull red block rests as still as marble statuary. A starry sky hangs just as still behind it, falling away untwinkling toward the infinite depths like a blizzard of distant electric lights. The monolith itself is deeply chilled and frosted—its depressions glisten with ice. Its upper surface is varnished with coatings of carbon, but there is not enough light to show their streaks of ochre and rust—the darkness here is deeper than the darkest desert midnight.

This aged edifice is not yet part of any landscape, icy or rocky—though that is its destiny. This is still a free-flying asteroid. The asteroids have been gradually disappearing over the eons, destroyed in the process of solar system construction. This is one of the last, having escaped destruction longer than most.

Now the red asteroid floats suspended, nearly stopping at the outermost reaches of its deep, elliptical orbit. It moves its slowest here at the turnaround point on a solitary sojourn—a place it has not been back to visit for thousands of years. It has been on farther-flung voyages than this—on orbits ten thousand years long and longer. On swings as deep as those, the stars have time to shift around in their constellations—altering their patterns from one orbit to the next.

The starry points that always changed the most behind the asteroid's silent voyage were the ones back toward the brightest light in its night—the sun. Those particular wandering lights were uniquely colorful—some glowing blue-green, some golden, or reddish, or shades of ivory. As the asteroid receded toward the farthest extremes of its orbit, all those colored stars moved inward to align with the sun; then they moved in closer still, finally disappearing in the sun's glare. On the asteroid's inbound flight, they reappeared. They all expanded out on their common axis, moving to both sides, some of them finally sliding behind the asteroid as it fell toward the center of the solar system.

Since its birth, the asteroid had weaved among those inner lights in the solar system—the planets—following an ever-changing track between them. Every so often, once in a hundred orbits or more, one of those lights would grow unusually bright, and its point of color would expand to a disk—a crescent from behind, a full circle from the sun side. During such an encounter the gravitational attraction of the approaching world curved the path of the asteroid toward it—the closer the pass, the greater the effect. The result would be a new orbit in which the asteroid's inbound course fell nearer the sun, gradually vaporizing its skein of accumulated ice.

But if it passed in front of a rapidly approaching planet—so that the greatest gravitational pull came from behind the asteroid at closest approach—the asteroid's forward motion would be reined back. This deceleration would shift its track deeper still beyond the planets, increasing its time spent in the frigid outer reaches. There it would become covered in a frosting of carbon soot and ice, swept up one molecule at a time over the cold millennia.

An orbit that flies between the planets must be special for its occupant to survive the eons. Planets are asteroid killers—if orbits intersect. The orbits that support most of the remaining asteroids avoid the planets by curving above and beyond their plane, or never crossing their tracks for some other reason; lunar-capture orbits also preserve asteroids—temporarily.

But orbits change. The asteroids survive only by chance, weaving among the larger spheres on constantly evolving trajectories until they finally wander too close to the irresistible gravity of some planet or its moon or the sun. The safest orbits belong to the outer-system asteroids. Over their long lives, they accumulate thick shells of frozen gas, sweeping up smaller bits and pieces as they drift ever more slowly on their timeless passages—following orbits so far away that they never come closer to the sun than Pluto.

Pluto is an example of a small body that follows a planet-crossing orbit but survives nonetheless. It is an icy planetoid from the edge of the outer-system disk of asteroids. Its orbit has come to cross that of Neptune, and it survives only because its track happened to synchronize with that of the larger body. Now when it intersects Neptune's path, the synchrony of the orbits always finds Neptune in one of two specific spots, each equally far away from the crossing point.

There is not much chance of an encounter out there—beyond the major planets—to dislodge Pluto from its schedule. This track is stable. But the paths of small bodies are always accumulating random perturbations;

Deep space. The asteroid may spend
a thousand years beyond Pluto on its deepest orbits.

eventually the changes move them out of their secure routines. Some
chance impact may eventually nudge Pluto away from its synchrony with
Neptune. Then over the millennia the odds would grow that one day Neptune would be right in front of Pluto at the crossing.

And then, as has happened so many times with similar objects in
Neptune-crossing orbits, there would be a titanic explosion. The sea green
clouds that shroud the planet would roil and darken, and when the atmosphere finally settled, the rocks and ice of Pluto would be no more than a
small part of a larger planetary body in a slightly altered orbit.

A similar fate awaits Triton, a body similar to, but a bit larger than,
Pluto. Triton has already been captured by Neptune. It is now in an unstable lunar-capture orbit, spiraling gradually inward toward the giant green
planet. It will eventually be ripped to pieces as it closes in on the cloud tops
below it, ending its existence as a Saturn-like ring around Neptune, tinted
sea green in reflected planet light.

Phobos is another example of an asteroid in an unstable orbit. Phobos is a twelve-mile-wide rock-and-iron asteroid that was jostled out of its circular track inside the orbit of Jupiter and wandered too close to Mars. But its uncommon convergence with the red planet postponed its demise. As Phobos fell toward the larger body, the arc of its path overshot its impact. The asteroid swung over the curve of the far horizon but could not swing free—it remained held in the grip of the Martian gravity. Its lunar-capture orbit is descending—faster with each pass. By now it is only a few thousand miles above the surface. In some millions of years, it will finally disappear beneath a planetwide sandstorm raised by the excavation of a chain of craters splashed across the red desert by its own destruction.

The red asteroid is more than four billion years old. When it was created, it flew in the company of millions of other asteroids. Many of them orbited in a disk beyond the orbit of the planet farthest from the sun—Uranus. From the Earth, Uranus was easily visible, as was the brighter Neptune. Their unique green lusters stood in contrast to the warmer colors of the stars and the other planets that populated the night.

Back then, the planets were growing—through a succession of asteroid impacts. Asteroids were the ferries of that time. As they followed their ever-changing orbits, they carried water (in its frozen form) inward toward the sun—depositing it as oceans or ice caps when they chanced to land on the inner planets. Others of them ferried iron and rock outward, adding their mass to that of the outer planets they ran into.

Asteroids ferried not only mass around the solar system, but also energy. All the bodies in the solar system—be they planets or pebbles— constantly pull on each other with their gravitational attraction. The closer they pass, the more strongly their presence is felt; farther away, the less effect their pull has on trajectory. When energy is exchanged between two bodies, speed is increased in one, leaving the other coasting commensurately slower. If asteroids encounter planets, their course and speed change much more than do those of the larger body. The accelerated asteroid carries its added energy for millions of years, losing no velocity to its frictionless passage through space. Its speed will be conserved until it again gains or loses energy in its next close encounter.

The surviving asteroid now approaches Jupiter, speeding inbound on its elliptical orbit. It had been accelerating for months, falling faster each day into the giant planet's gravity well. Bands of storm rotating around the planet become clearer as the massive disk grows to fill the sky dead ahead of the asteroid.

Jupiter finally shows some lateral motion only with the asteroid's closest approach. The giant planet moves beneath the asteroid, even though the smaller body's course is curving down ever more sharply toward it.

The asteroid bends around Jupiter, the speed of the added gravitational acceleration carrying it clear. But at closest approach it catches the uppermost edge of the atmosphere. Aerodynamic resistance with the cloud tops deflects the asteroid down into thicker clouds, where air friction converts its speed into heat, slowing its passage and giving the planet more time to pull it closer.

As it scorches the cloud tops, the plummeting missile burns a path through the dark-side Jovian sky like a coal-fired locomotive barreling through the fog at midnight. Its fires illuminate the cloud of steam that envelops it; its trains of sparks and smoke diminish in brilliance as they shrink away behind.

The asteroid's surface melts in the friction of its passage. Its speed compresses the atmosphere ahead of it into a wall harder than the asteroid itself—against which the asteroid fractures. A chunk of one side breaks off to plunge below, deflecting the remainder upward.

The escaping pieces emerge through the clouds, trailing red banners of ionized hydrogen above the night side of the planet. The largest piece whirls madly, most of its surface glowing sun white but with a cooler, internal facet showing dark through a newly fractured face. The alternating bright and dark areas cause the spinning asteroid to blink, as seen from afar.

Having left much of its mass and energy with Jupiter, the asteroid now coasts away on a new trajectory that aims it closer toward the sun. Behind it, the dark side of the receding giant rotates into the light. The emerging hemisphere of clouds is divided by a black trail of burnt atmosphere that gradually breaks into dashes—dark streaks carried apart by the counter-rotating bands of storm they cut across.

In the earlier years of the solar system, some of the asteroids survived because their close encounters flung them clear of the paths of the planets. These asteroids came to reside in a ring beyond the farthest planetary orbit. But the transfer of energy between the major planets and the millions of asteroids eventually moved the planetary orbits into conflict with each other. The result was chaos.

The giant planets came close enough to each other to exchange massive amounts of energy directly—throwing each other into elliptical orbits. Those orbits widened until they drifted out toward the ring of asteroids. Then the smaller bodies were again pulled from their courses and flung around the solar system.

On one of its excursions, Neptune passed Uranus and ventured deep into the outer asteroid disk. It began to sling asteroids inward toward the sun, which had the effect of moving the planet itself farther outward into a more circular orbit. It eventually receded to twice its previous orbital radius, and its greenish glimmer faded to invisibility from the skies of the inner solar system.

The asteroids flung inward during this epoch carried their mass and energy back into the inner solar system—renewing the process of planet building. Once in a century, one of them would slam into one of the inner planets. The face of the moon is a record of that era of bombardment, which lasted millions of years.

Stability eventually returned to the planetary orbits. The larger bodies fell into gravitational resonance with each other, each one stabilizing the others. Their continual energy transfers with the remaining asteroids smoothed their elliptical orbits into circles.

Millions of years after the Jupiter encounter, the wandering asteroid once again feels the attraction of a planet, this time dead ahead. After days of steady gravitational acceleration, the asteroid passes through a thin atmosphere, vaporizing its own crust of ice in seconds. It does not slow until it passes between parallel mountain ranges and through the surface of a forested plain. It plunges a quarter mile into the ground before its penetration is blocked by the bedrock, decelerating from planetary orbital velocity to zero in a hundredth of a second.

In that instant, the energy of its thousands of tons of momentum is converted from speed to heat—melting the entire asteroid, boiling the melt to vapor, and ionizing the vapor to a plasma hotter than the interior of the sun. The buried, incandescent cloud finds release along the path of least resistance, erupting straight up through the surface. So even though the angle of impact was oblique, the crater shaped by the subterranean explosion will not be oblong but as circular as if the strike had been vertical.

The force of the explosion transfers the asteroid's energy to the larger body, altering the planetary orbit so slightly that we could not measure the change. The detonation obliterates everything in a three-kilometer radius from the impact site, torching the forest out to six kilometers, felling trees and maiming animals out to ten. Shock-wave winds diminish to hurricane force twenty-five kilometers from the center. Rock and iron vapor condense from the sky, adding the mass of the asteroid to the Earth in a rain of dust and glassy beads across the valley and far beyond.

The largest pieces of fractured rubble tumble back to fill in most of the crater. Smaller fragments of bedrock hurl away at supersonic speeds, throwing shards of ground zero up through the atmosphere and into unstable orbits that soon spiral back down far across the globe. Some of that material accelerates to escape velocity—bits and pieces of sandstone fly away from Earth to replace the asteroid in orbit around the sun.

An incandescent pool of slag in the bottom of the crater remains as the focal point of the decimated valley. The molten lake will take centuries to cool. The broad, sterile impact scar will eventually be assimilated into the landscape, as such scars have always been and will continue to be.

The sandstone rocks ejected from Earth by the impact are foreign to the other minerals in solar orbit (the process of sandstone formation takes place underwater). But the grains in the newly orbiting stone, even the bacterial spores deep within its cracks, are all ultimately made of material that initially solidified in solar orbit and was long ago delivered to Earth by impact from space.

The planets are still growing. Earth accretes tons of mass every day under the rain of particles from space, an infall that lessens as the eons lengthen. But most of the era of asteroid-mediated construction and destruction is behind us, and the solar system is stable. It could have turned out differently.

The giant planets could have collided, slowing one of their members until it wandered into the inner solar system, where it would fling the smaller planets away into space with its massive gravity, eventually falling into the sun itself. Our system could have matured with fewer or no planets.

Such chaos seems to have been the rule for other planetary systems. There, gas giant planets that formed far from their central stars have transferred their orbital energies during close encounters, slinging their smaller siblings away from warm, near-sun orbits into the absolute cold of interstellar space. The planets that did the slinging have spiraled down into orbits so close to their central stars that they now evaporate in the blaze of their own sun's stellar wind.

Ours may be a rare condition—where large numbers of planets formed, together with larger numbers of asteroids to act as liaisons to the process of orbital stabilization. The asteroids transfer energy gradually, in multiple small increments between the bigger bodies. This facilitates the circularization of the orbits of the surviving spheres, and their distribution uniformly across the plane of the system.

In that scenario, one or two planets will be left orbiting in that comfort zone where water is neither frozen as hard as granite nor evaporated immediately to space. And there the stage is set for the establishment of life.

When the residents of habitable worlds finally develop the capacity to cast their gaze skyward, they will always find multiple planets passing among their stars—anchors of the process that stabilized their home world in an orbit where life could evolve.

Science Notes

Interactions between pieces of rock and ice large and small over the history of the solar system can be viewed as constructive (planet formation, orbital circularization) or destructive (impact cratering, planet ejection from the system). A Mars-sized planet was destroyed during the genesis of the Earth, but the moon was created as a byproduct (Stevenson, 1987). Planetary bombardment would have sterilized planets, the molten impact craters overlapping to melt the entire surface and boil away oceans. But for the sustenance of life, the end result of all the interactions of our system seems to have been positive. Some other solar systems appear to have lost most of their planets—the smaller ones flung away into free space during gravitational close encounters with the largest ones, which then migrate inward toward their suns (Murray et al., 1998). In the case of our system, the interactions between all the pieces of matter in orbit around the sun, from the tiniest grains to the largest planets, have eventually produced a series of concentric, nonconflicting orbits.

In this tale, all the subplanet-sized objects are called asteroids. During the early epochs, gravitational interactions with asteroids altered the orbits of the planets (Murray et al., 1998). Planets can move outward as they throw asteroids inward (Strom et al., 2005). Their final orbits are also determined by their interactions with each other (Thommes et al., 2002). Planets in our system have traded places as a result of these interactions. Uranus was once closer than it is now (it is now visible to the naked eye, at magnitude +5.8, so it would have been even brighter then, as seen from Earth), and Neptune was even closer to the sun than Uranus before they switched (Gomes et al., 2005; Tsigranis et al., 2005); now, Neptune is no longer naked-eye visible from Earth.

Planets are built by sweeping up the asteroids that cross their paths, through a series of collisions (Schilling, 1999), a process that is ongoing (Wynn and Shoemaker, 1999). Near misses deflect the asteroids into new orbital trajectories. The "atmospheric skip" near miss of the asteroid passing Jupiter in this tale is analogous to the skimming pass of an asteroid across the top of Earth's atmosphere in 1972 (Rawcliffe et al., 1972). Though Pluto is in a planet-crossing orbit, its 2:3 orbital resonance with Neptune is stabilized by a continuous gravitational interaction between the two bodies (Malhotra, 1993); the resonance leaves Neptune in one of two alternating locations (both far from the crossing point) when Pluto crosses its orbit. Some asteroids fall into lunar-capture orbits before their eventual impact (Showalter and Lissauer, 2006). There are many modes postulated whereby an asteroid can be so captured (Kuiper, 1956; Jewitt and Sheppard, 2005), including collisional deceleration, atmospheric breaking, and transfer of one member from a rotating, gravitationally bound asteroid pair (Angor and Hamilton, 2006).

An asteroid passes through the four states of matter (solid, liquid, gas, and plasma) during the instant of impact (Melosh, 1989). Plasma is the state of matter in which atoms are torn apart and electrons are separated from their nuclei to fly free; this state of matter is a radiant vapor. Escape velocity is that speed at which a body's course cannot be bent back into an orbit by the gravity of the bigger body from which it is departing. Moons reaching that velocity depart from their planets, as do planets from their suns.

References

Angor, C. B., and D. P. Hamilton. 2006. Neptune's capture of its moon Triton in a binary-planet gravitational encounter. *Nature* 441:192–94.

Gomes, R., et al. 2005. Origin of the cataclysmic late heavy bombardment period of the terrestrial planets. *Nature* 435:446–49.

Jewitt, D., and S. Sheppard. 2005. Irregular satellites in the context of planet formation. *Space Science Review* 116:441–55.

Kuiper, G. P. 1956. On the origin of the satellites and the Trojans. *Vistas in Astronomy* 2:1631–66.

Malhotra, R. 1993. The origin of Pluto's peculiar orbit. *Nature* 365:819–21.

Melosh, H. J. 1989. *Impact cratering: A geologic process*. New York: Oxford University Press.

Murray, N., et al. 1998. Migrating planets. *Science* 279:69–72.

Rawcliffe, R. D., et al. 1972. Meteor of August 10, 1972. *Nature* 247:449–50.

Schilling, G. 1999. From a swirl of dust, a planet is born. *Science* 286:68.

Showalter, M. R., and J. J. Lissauer. 2006. The second ring-moon system of Uranus: Discovery and dynamics. *Science* 311:973–77.

Stevenson, D. 1987. Origin of the moon: The collision hypothesis. *Annual Review of Earth Planetary Science* 15:271–315.

Strom, R. G., et al. 2005. The origin of planetary impactors in the inner solar system. *Review of Earth Planetary Science* 309:1847–50.

Thommes, E., et al. 2002. The formation of Uranus and Neptune among Jupiter and Saturn. *Astronomical Journal* 123:2862.

Tsiganis, K., et al. 2005. Origin of the orbital architecture of the giant planets of the solar system. *Nature* 435:459–69.

Wynn, J., and E. Shoemaker. 1999. The day the sands caught fire. *Scientific American* 279:36–45.

A Dangerous Place

These days, we are passing through an exceptional diorama. Just now, our ride on the Earth—along the track the sun follows through space—is taking us across one of the spiral arms of our galaxy. Those arms are the namesake features of grand spiral galaxies, such as our own Milky Way.

Seen from a perspective a hundred thousand light years above, set against the void of deep space, our galaxy is dominated by those great curved spokes. Lit with clusters of blue points of starlight and woven with red fluorescent clouds, they arch away across the massed brilliance of the billions of stars in the galactic disk. From that high vantage point, our sun is an inconsequential mote, invisible without a telescope.

The vast, curving arms suggest that spiral galaxies are great pinwheels whose huge momentum holds their rotation down to just below the speed of perceptible movement. The illusion is reinforced where the ends of the spirals are bent back, as if by greater centrifugal forces at their faster-moving outer edges.

In fact, any analogy with the familiar motion of a solid disk is misleading. Stars flow through galaxies not in unison but at different speeds—slower the farther they are from the center. Our sun takes hundreds of millions of years to complete one trip around the great circle. And the spiral arms themselves advance even more slowly. They are standing waves in a sea of light, continually falling behind the individual stars, which then pass right on through them.

Now that we are in one of the arms, we have a sky full of brighter-than-usual lights to gaze upon at night—since the density of stars within the arms is greater than it is between them. But we are too close to the arm itself to see it. Only from a distant remove would its bright young stars and glowing clouds of gas shrink together and coalesce into a river of light streaming away from the galactic center.

If we could spend a while gazing at a night sky in which a thousand years pass each second, we would be captivated by the stellar motion. The shimmering river of the Milky Way (made up of the galaxy's more distant arms) would remain steadfast overhead, but we would see the nearby stars wan-

Day star. A supernova explosion bright enough to see by day lights the dark side of the moon at the end of the Permian period.

der from their constellations to continuously rearrange themselves into new figures for us to name, dissolving from one grouping into another.

Stars tracking toward us would brighten—the larger of them coming to dominate the scene. Should we move into a star cluster—a stellar nursery full of new, young stars—we would be treated to a sky so crowded that our atmosphere, seen between the stars, would lighten at night from black to deep sapphire blue.

As our imaginary night lengthened, most of the stars would dim and diffuse away. The night sky would lose its wonder as Earth's orbit through the galaxy carried us into the safer, emptier quarters between the spiral arms.

When the orbit of our sun through the galaxy leaves us between arms, the night sky presents a darker, lonelier vista. But the arc of sparkling stars that now dominates our evenings has a downside. Spiral arms can be dangerous places to visit.

We are not as safe while we are within a spiral arm, because there—where the density of stars is greater—the odds of a close encounter with a star are greater. Within an arm, the chances increase that a blue supergiant star like Rigel or Deneb might come as close to the sun as Sirius is now. Such a passing giant would shine with the brightness of the moon concentrated into a single star point—250 times brighter than Venus. Its actinic blue spark would cast a shadow at night and be visible against the sky all day long. But even that close, the powerful ultraviolet radiation of such a star—or its fierce stellar wind of charged particles—would pose no threat to Earth. We would be insulated by the intervening gulf of light years.

Isolation through distance protects us from every threat such a star would offer, with one exception. The Earth would come to great danger should such a star explode nearby—and supergiant stars are prone to explosion. Should the life of a supergiant star end in a supernova while it was passing us, the consequences would be devastating. A nearby supernova would create a second sun in our sky—a violet white beacon that would be painful—in fact, blinding—to look at. The explosion would change the chemistry of the air, which would mean that the nova could beam its intense ultraviolet light straight through the atmosphere into our eyes. The daytime sky would darken from blue to orange, and at night it would shine with the shifting pastel phosphorescence of a worldwide aurora.

The nova would dim over the ensuing months, leaving an expanding globe of twisted, glowing wisps of gas in its place in the night. But the bombardment of our atmosphere by the shrapnel of high-energy particles from the explosion would continue for hundreds of years.

Today, a hole in our atmospheric ozone layer exposes the surface of the Earth around the poles to the damaging ultraviolet portion of the sun's radiation; the rest of the planet is protected from those biocidal rays where that layer remains intact. But the bombardment of Earth by the fallout from a nearby supernova explosion would expand that ozone hole—to expose all of Earth's surface to the ultraviolet radiation generated by our own sun, for centuries. The continuous exposure would threaten the biosphere, debilitating the food chain from the bottom up.

Even if it does not explode, a star of any size passing close by the Earth can have an impact here in another way. Its gravity competes with the influence of the sun's distant pull to alter the orbits of the most far-flung of the millions of chunks of rock and ice that make up the Oort cloud. This halo of debris left over from planet formation extends the solar system to almost a light year in diameter.

Some of these chunks may have their paths stretched so that they fly through the inner solar system on one leg of an altered orbital track, raising the possibility of planetary impacts. Such a bombardment of the Earth by comets or rocky impactors resulting from the passage of a star close by the Oort cloud is more likely while we are within a spiral arm, where the stellar density is higher.

Now that we are within a spiral arm, the view around us is inspiring indeed. Spiral arms are sites of star birth. The arms are shaped by pressure waves, where bursts of intense starlight push veils of hydrogen gas ahead of them. Every kind of cosmic radiation roils this maelstrom—including powerful solar winds from blazing young stars, and shock waves from supernova explosions.

Sheets of hydrogen compressed from all sides by these shock winds achieve a density at which they begin to collapse under their own gravity into star-forming clouds. These knots of gas contract until they are blown apart again by the heat of new stars of all sizes igniting within them. The spiral arms shine in the intense blue light of the largest of those newborn stars.

Our arm is named for one of the constellations it lies behind, from our perspective—it is the Orion arm of the galaxy. It contains the brilliant stars of the winter constellations that are now predominant in our sky. Its star-forming clouds also lie close all around us. Interstellar plains of hydrogen float above in every direction—stretched immensely wide and thin, glimmering with a red fluorescence too faint for us to perceive with the unaided eye.

In places, these star-forming clouds have been compressed enough to become visible. The fuzzy, central "star" we see in the "sword" in the constellation of Orion is actually a glowing whirlpool of hydrogen many light years across. It is set alight by the radiance of the cluster of new stars being born within it.

All around our sky, from the celestial pole down into the southern hemisphere, other star-forming nebulae glow in the Orion spiral arm. The space between them is populated by groupings of young stars that formed from earlier clouds—groups such as the Pleiades or the Beehive Cluster. Those stars still remain together in the area where they were created—though the pressure of their starlight has now dispelled the nebulae in which they formed.

The longevity of a young star is inversely proportional to its size. Our sun, a medium-sized yellow star, will live for billions of years. The largest, brightest stars born in the spiral arms are the supergiants. They die in only millions of years, exploding before they have time to pass beyond the arms where they were born. They are found only within spiral arms, and so supernovae that mark the ends of their existences occur only within the arms as well.

The evolution of life on Earth is chronicled in the fossil record. New taxa evolve while older ones disappear, leaving nothing more than their imprints in the rocks. Those disappearances of species from the Earth seem to be linked in time.

Geologic strata of stone exposed on a cliff face may each be only millimeters thick, but still they represent thousands of years of geologic history. Some of these strata appear as barriers below which fossils of whole groups of ancient plants and animals predominate but above which they are all absent. These linked disappearances of species from one sedimentary stratum to the next reflect catastrophic events in ancient times—mass extinctions.

Mass extinctions in Earth's flora and fauna may be due to climate change—periods of temperature extremes, such as global warming or cooling events that led to ice ages. Mass extinctions have also been ascribed to impacts of icy comets or rocky debris falling from space. Extinctions are more difficult to ascribe to nearby supernovae, because the alteration of the atmosphere leaves no prominent geological signature in the rocks.

A succession of mass extinctions has marked the history of life on Earth. When astronomers extrapolate the orbit of the sun back across those times, they find that the dates of those extinction events correlate with times when the Earth was crossing one of the galactic spiral arms.

For instance, the trilobites had been the predominant class of creatures in the Cambrian seas for 90 million years when the sun entered the Norma arm of the galaxy 450 million years ago. Then the world these creatures knew changed. The food chain they depended on altered more rapidly than they could adapt—bringing the age of their dominance to a close.

Most of the creatures in the shallows—the parts of the ocean most directly exposed to the sky—died off. And then the world cooled, plunging into a protracted ice age from which the trilobites never recovered. Their few remaining species survived until the Permian mass extinction two hundred million years later. Then the sun entered the Crux arm of the galaxy, and the remaining trilobites disappeared forever along with 95 percent of all marine species extant at that time.

Most of the species the world has ever known have gone extinct—most of them during mass extinctions. There have been five major mass extinctions since the days of the trilobites. The last one, ending the Cretaceous period, came sixty-five million years ago, when the sun had entered the Sagittarius arm of the galaxy. All land animals greater than twenty-five kilograms in weight went extinct at that time.

Now, just as the epoch of the predominance of our own species has begun, we begin the trek across one more galactic spiral arm. The stars have brightened around us with our passage away from the calm space between arms. Nearby stars will be wandering closer over the next million years. We will be wondering at the brightest of them as we look out into the evening sky and ponder our place in time and space.

Science Notes

The sun is 8,000 parsecs (1 parsec = 3.26 light years) from the galactic center. (Astronomers are now shifting from light years to parsecs as a measure of great distances.) Traveling at 220 kilometers/second, it takes the sun two hundred million years to make one circuit around the galaxy. There have been twenty such circuits since the sun's birth. Ours has been a barren world most of that time. There has been life enough to register extinctions in the fossil record here for only the past two and a half sun orbits around the galactic center. Each cycle carries the Earth through the spiral arms, during which passage the danger to the biosphere heightens. The historical order of arm crossings listed here is from Leitch and Vasisht, 1998; however, uncertainties about the actual positions of the arms, about arm windings, and about the constancy of the sun's orbit leave extrapolations of the positions of the arms through time uncertain (cf. Gies and Helsel, 2005).

There will always be debate about the factors contributing to mass extinctions, since we cannot travel back in time and resolve the hypotheses. Bombardment by comets may increase during arm passage (Hills, 1981). Supernova irradiation may upset atmospheric chemistry, leading to extinctions (Ruderman, 1974; Cockell, 1999; Benitez et al., 2002; debated by Gehrels et al., 2003). Intense irradiation of the atmosphere may lead to cooling of the Earth through the production of atmospheric nitrogen dioxide, which absorbs blue light and therefore decreases the solar radiation reaching the surface (and changes the sky color). Nitrogen oxides catalyze the destruction of ozone. A general increase in cosmic radiation, as is encountered within spiral arms, could lead to global cooling by increasing ionization in the atmosphere. This would lead to increased charged aerosols that promote the formation of condensation nuclei, increasing cloud cover—which reflects solar energy back into space, shading the surface until it cools (citations in Gies and Helsel, 2005). Mammals may die under an ozone-depleted sky from vitamin-K poisoning; the final step in synthesis of that vitamin takes place in skin, in a reaction catalyzed by and proportional to the available level of sunlight. Terrestrial causes—such as episodes of flood basalt volcanism—are also potential contributors to global extinction of species.

Mass extinctions have been identified in the fossil record at the ends of geologic periods. The bigger ones occurred at the end Cretaceous (65 million years ago [mya]), the end Triassic (200 mya), the end Permian (250 mya), the end Devonian (360 mya), and the end Ordovician (444 mya) (Raup and Sepkoski, 1982). The Permian event appears to have been particularly harsh, killing off 70 percent of the land species. It was the only extinction to kill off a significant fraction of the extant insect fauna; the fossil record at that time shows a coal gap, which is interpreted to reflect an extensive decrease in plant cover across the Earth; that record also shows a spike in fungal growth during the several millions of years after the event. All the mass extinctions could have been caused, entirely or in part, by extraterrestrial factors that led to changes in the atmosphere or impacts on the surface or both.

The star-birth theory of spiral arms (Gerola and Seiden, 1978) suggests that if you could float along with the galaxy's stars, moving forward through time at their speed so that they appeared motionless below you, then the glow of a passing spiral arm would move backwards—in the direction opposite to star motion. New stars would be born to provide the extra light that makes the arms stand out. Those new stars would move at the speed of the other stars already populating the disk—born from gas clouds moving at that same speed. What would move backwards is the wave of compression that initially fired that gas, instigating star birth. That pressure comes from the excessive starlight of hot stars newly born in the spiral arms, and from shock waves from the largest members of those newborns that exploded soon after their birth. The pressure of that starlight and those shocks compress the gas clouds in their midst into more new stars (Draine and McKee, 1993; Vishniac, 1994). The new stars, especially the giants, produce more starlight pressure—propagating the reaction around the disk.

The galaxy's four major arms form from two spokes that cross through the galactic center. Each arm takes more than two hundred million years to make one circuit around the disk. Blue supergiants are born and then explode in a span of two million years or less, so they are found only in the spiral arms. Arms are named for the regions in our night sky that they dominate: the Sagittarius-Carina arm, the Scutum-Crux arm, the Norma-Cygnus arm, and the Perseus arm. We are now entering a smaller spur—an arm segment between the Perseus and Sagittarius-Carina arms called the Orion arm. That arm contains the bright stars of the northern hemisphere winter sky and star clusters such as the Pleiades (in Taurus) and the Beehive Cluster (in Cancer). Rigel (in Orion) and Deneb (in Cygnus) are intensely bright blue supergiants—the most distant first-magnitude stars in our skies. If one of them were passing as close to us as Sirius (2.64 parsecs), it would shine as a point source of magnitude −10; the full moon is magnitude −12. Should one of those supergiant stars explode in a supernova, its light output could temporarily increase a billionfold.

References

Benitez, N., et al. 2002. Evidence for nearby supernova explosions. *Physical Review Letters* 88:081101.

Cockell, S. C. 1999. Crises and extinction in the fossil record: A role for ultraviolet radiation? *Paleobiology* 25:212–25.

Draine, B. T., and C. F. McKee. 1993. Theory of interstellar shocks. *Annual Review of Astronomy and Astrophysics* 31:373–432.

Gehrels, N., et al. 2003. Ozone depletion from nearby supernovae. *Astrophysical Journal* 585:1169–76.

Gerola, H., and P. E. Seiden. 1978. Stocastic star formation and spiral structure of galaxies. *Astrophysical Journal* 233:129–35.

Gies, D. R., and J. W. Helsel. 2005. Ice age epochs and the sun's passage through the galaxy. *Astrophysical Journal* 626:844–48.

Hills, J. G. 1981. Comet showers and the steady state infall of comets from the Oort cloud. *Astrophysical Journal* 86:1730–40.

Leitch, E. M., and G. Vasisht. 1998. Mass extinctions and the sun's encounters with spiral arms. *New Astronomy* 3:51–56.

Raup, D. M., and J. R. Sepkoski. 1982. Mass extinctions and the marine fossil record. *Science* 215:4539–43.

Ruderman, M. 1974. Possible consequences of nearby supernova explosions for atmospheric ozone and terrestrial life. *Science* 184:1079–81.

Vishniac, E. T. 1994. Non-linear instabilities in shock-bound slabs. *Astrophysical Journal* 428:186–208.

Follow the Threads on the Web

The Web sites in the lists that follow contain information on natural history, including material that describes and expands upon many of the topics covered in this volume. Any of these well-produced resources will take as long to appreciate as a good book takes to read. Many present animations, videos, and even live videocams, as well as further links. These sites contain a range of different approaches to the presentation of the observations, data, and theories of natural science.

The Physical Environment

1. Atmospheric optics. Visions from the clear sky, by Les Cowley. *www.atoptics.co.uk*. An inclusive treatment of phenomena of the sky, illustrated with fine examples of each, and explained in detail.
2. Basic meteorology, from the Department of Atmospheric Sciences at the University of Illinois at Urbana-Champaign. *www2010.atmos.uiuc.edu/(Gh)/guides/home.rxml*. Introduction to the weather. See also the introduction to severe weather: *severewx.atmos.uiuc.edu/index.html*.
3. Basic volcanology, assembled by a team of Swiss and Italian scientists and educators. *www.swisseduc.ch/stromboli/index-en.html*. The site offers a comprehensive image collection, geologic background, and virtual tours.
4. Tectonics. Presented by the educational multimedia visualization center of the Department of Earth Sciences at the University of California, Santa Barbara. *emvc.geol.ucsb.edu/index.htm*. Descriptions of the dynamic surface of the Earth, with a subemphasis in western North America, e.g., around Santa Barbara.
5. Geologic time. From the Smithsonian and the National Museum of Natural History. *paleobiology.si.edu/geotime/main/index.html*. An introduction to the paleobiologic eras of the Earth.
6. Cloud physics. The Web site of Kenneth Libbrecht, chair of the Physics Department at Caltech. *snowcrystals.com*. On the science and the beauty of snowflakes.

The Biosphere

7. Monterey Bay Aquarium and the Center for the Future of the Oceans. *www.montereybayaquarium.org.* A resource of wildlife and conservation information about the sea and shore.

8. Wayne's Word. An online textbook of natural history, by W. P. Armstrong, used with his courses in the Life Sciences Department at Palomar College. *waynesword.palomar.edu/indxwayn.htm.* A collection of natural histories with an emphasis on plants of the American Southwest.

9. e-Nature. Originally managed by the National Wildlife Federation. *www.enature.com/home/.* Online field guide to common wildlife species.

10. Microbiological garden. From the Institute of Chemistry and Biology of the Marine Environment at the University of Oldenburg, Germany. *www.icbm.de/pmbio/mikrobiologischer-garten/eng/index.php3.* An introduction to a most complex aspect of the natural world—the world that opens up through the microscope.

11. Trilobites. A Web site created by Sam Gon III. *www.trilobites.info/.* Everything you've ever wanted to know about this extinct order.

The Rest of the Universe

12. Atlas of the Universe. Created by Richard Powell. *www.atlasoftheuniverse.com/index.html.* An expandable map of Earth's celestial environment, with a hyperlinked glossary.

13. NASA. *www.nasa.gov/home/index.html.* A searchable site featuring current news of space exploration, an emphasis on education, and other features, including a planetarium-style tour of tonight's sky. See also *imagine.gsfc.nasa.gov/* and *space.jpl.nasa.gov/.*

Physics and Math

14. Exploratorium, the museum of science, art, and human perception, San Francisco. *www.exploratorium.edu/explore/index.html.* An interactive environment dedicated to the understanding of physical phenomena, with an emphasis on the way things are perceived.

15. Fibonacci numbers and the golden section. By Dr. Ron Knott, hosted by the Mathematics Department of the University of Surrey, U.K. *www.mcs.surrey.ac.uk/Personal/R.Knott/Fibonacci/fib.html.* A description of the calculable basis of the nautilus shell spiral, the arrangements in seed heads, leaves on stems, and more.

Index

Alaskan yellow cedar, 148
albatross, 101, 108, 109, 125
alder, 18, 34
American white pine, 63
anchovies, 128
Andes Mountains, 172
anemones, 95
anglewing butterfly, 155–57
Antarctic convergence, 107
anthicid beetles, 134, 135
ants, 159
aposematic coloration, 137
aspen, 62, 144
asteroid, 189–96
aurora, 85

basilisk lizard, 53, 55
bears, 66, 115
beaver, 22–29
Beehive Cluster, 202, 203
behavioral parasites, 159
Beringia, 63
Bernoulli's Principle, 74
Bighorn Mountains, 65
Bitterroot Mountains, 60
black oak, 15–20
blister beetle, 133
blue supergiant star, 200, 205
bombardier beetle, 14
boobie, 101, 125–29
boring beetle, 146, 165
box jellyfish, 94

bristles, 44, 158, 159
brown creeper, 154–57
buckwheat, 16, 19

cambium, 147
Cambrian period, 203
Canidae, 14
Canis Major, 170
Canis Minor, 170
Canopus, 89
cantharidin, 134, 135
Carboniferous period, 42
Carnivora, 14
Cascade Mountains, 32, 67
chestnut blight, 64
chipmunk, 145, 146
cinnabar, 172
circumzenith arc, 185
Clark's nutcracker, 60–66
click beetle, 148
cnidarians, 99
coal gap, 204
cobra lily, 32, 34, 35
copper, 172, 175
corals, 96
cornea, 170
coyote, 10, 12–14, 23–26, 139, 142
crayfish, 6, 25, 41
Cretaceous period, 203, 204
crocodile, 95
cross-quarter day, 166, 167
cryptic coloration, 157

ctenophores, 93, 97, 99, 100
cubozoans, 98
curlews, 121

Darwin, Charles, 47, 55
dawn redwood, 18, 66
Deneb, 200, 205
Devonian period, 204
diamond dust, 83
diffraction, 43
Dog Star, 165, 166, 167
dolphin, 102–5, 108
Douglas fir, 148, 154
dowitcher, 87
dragonfly, 6, 39–42
dunlins, 87

eagle, 3–8
echinoderms, 98, 99
elk, 22, 142, 143
Europa, 165

false killer whale, 108
fire chaparral, 15
fire pines, 59
firefly, 135, 136
fireweed, 58
five-needle pines, 63, 67
fly, 158, 160
flying fish, 105, 126
flying frog, 52, 53, 55
flying lizard, 52, 55
flying squid, 126
flying squirrel, 28, 145–51
Fomalhaut, 169
fool's gold, 176
frigatebird, 125–29
frog, 25, 50–52
fulmars, 108
funnel cloud, 71, 72

Gause's principle, 157
ghost pine, 139
ginkgo, 18, 66
godwits, 79
gold, 171–76
gooseberry, 34, 56, 64
grasshopper, 49, 50, 54, 55
gravity well, 192
Great Basin, 66, 67
green (sea) turtles, 94
ground sloth, 174
ground squirrel, 145, 146
gyrfalcon, 87, 88

Hawaii Isle, 177, 181, 182
heart rot, 147, 150
heart worm, 26
horses, 118, 142, 143
house cats, 142
humans, 140, 141, 143
humpback whale, 122, 128

ice age, 15, 202, 203
ice crystal, 47, 70, 84, 183–86
incense cedar, 16
indole, 10, 14
Isle Royale, 29

jacamar, 50
jaegers, 82
jet stream, 101, 169
Jupiter, 192–94

Kilauea Caldera, 179, 182
killdeer, 79
knots, 88
krill, 107, 125, 129

Labiatae, 14
ladybugs, 134

Lagomorpha, 14
Laplace's theorem, 169
lava flow, 178, 182
lava tube, 178–80, 182
lens, 166
Leopold, Aldo, 29
Lewis and Clark, 56, 60, 63–65
lodgepole pine, 7, 59, 63, 67
Loihi Isle, 181
lunar pillar, 84
lupine, 16, 19

mangroves, 128
manzanita, 15, 18, 20, 56
marbled cat, 52, 55
mare's tails, 186
margay, 49–55
marmot, 56, 57, 59, 62
Mars, 165, 175, 192, 196, 202–5
mass extinction, 202–5
Maui Isle, 181
Mauna Kea, 181
Mauna Loa, 179, 181, 182
Mayan civilization, 143
mayfly, 3–6, 8, 25, 45
menthol, 14
Mercury, 175
mica, 122
midge, 30, 45, 83
milkweed beetles, 134
mockingbird, 139, 142
mollusks, 98, 99
moon dogs, 84
moose, 22, 24, 174
mosquito, 6, 26, 28, 30, 40, 83
Mount Diablo, 122
Mustelidae, 14

Nashville warbler, 15
needlefish, 126

Neptune, 190–92, 194, 196
Norma-Cygnus arm, 203, 205
Nunavut, 81

Oort cloud, 201
optical fiber, 40
Ordovician period, 204
Orion arm, 205
osprey, 3–8, 32
ozone, 201, 204

paradise flying snake, 52, 53, 55
parhelia, 185
patagium, 145, 146
peregrine falcon, 69, 73–74
Permean period, 55, 203, 204
petrel, 101–9
phacilia, 16, 19
phalaropes, 87
Phobos, 191–92
photosynthesis, 150
pill bug, 158, 159
pillow lava, 182
pitcher plant, 30, 34, 35
placer deposits, 174, 175
Pleiades, 202, 203
plovers, 87
Pluto, 190–91, 196
Ponderosa pine, 16, 62
porbeagle shark, 99
porcupine, 10, 14, 28
prion petrel, 102
pupil, 40
pyrite, 122, 174, 176
pyrosomes, 93, 96, 98

quinone, 14

rabbit, 9–14, 139
rain shadow, 16

red queen theory, 54
refraction, 165, 168
remoras, 102
retina, 40
Rigel, 200, 205
right-eyed flounder, 112
Rocky Mountains, 64

Sagittarius-Carina arm, 203, 205
salps, 96
San Andreas Fault, 123, 124
sanderling, 78–90
Sawtooth Mountains, 64
Scutum-Crux arm, 203, 205
sea butterfly, 96, 99
sea wasps, 94
serotinus cones, 67
serpentine, 33
shark, 91–97, 104
shearwaters, 108
Sierra Nevada Mountains, 32, 34, 66, 67,
    172, 176
silversides, 128, 129
siphonophor, 92, 93, 99
Sirius, 89, 165–70, 200
skate, 114
skua, 80, 81
skunk, 10, 12–14
sky island, 18, 66
snowflake, 183, 186, 187
spider, 43–47, 93, 133, 134, 154
spiny-headed worm, 159
spotted owl, 148, 149, 151
stag beetle, 148
steelhead, 120
stone pine, 56, 60
sturgeon, 110–28
subduction, 176
sun dog, 186

supernova, 200, 201
surf bird, 87

tapeworm, 26
Titan, 165
tornado, 73–74, 101
Triassic period, 204
trilobite, 202, 203
Triton, 191
trophic cascade, 29
trout, 4–7, 27
truffles, 145, 151
tuna, 103–8, 126
tundra, 86, 87

Uranus, 192, 194, 196

Vega, 168, 170
Venus, 165, 168, 169, 175, 200
Vermillion Sea, 129

Wallace, Alfred Russel, 55
Wallace's flying frog, 55
Wasatch Mountains, 64
water strider, 24, 41
white-breasted nuthatch, 154–57
white-pine blister rust, 63, 64, 67
whitebark pine, 60–67
willet, 79, 105
willow, 23, 41, 62
wimbrels, 87
witches' broom, 146, 147, 150
wolf, 13, 23–29, 74, 116
Wollemi pine, 18
wood duck, 25

yellow warbler, 138–43
yellowlegs, 87
Yellowstone National Park, 29

Credits

Many of the images in this book were digitally created or composed by David DaRold. Thanks also to the photographers whose work appears on the following pages:

p. 37: 2006 © msphotoguy. Image from BigStockPhoto.com.

p. 45: Photograph by Nicky Davis.

p. 57: Photograph of Clark's nutcracker by M. Martin/VIREO.

p. 103: Photograph by G. Tapke/VIREO.

p. 178: 2006 © Lara Marieta. Image from BigStockPhoto.com.

p. 185: Photograph by Kenneth Libbrecht.

p. 191: Photograph of starscape by Paul Feldstein.